Technical Knowledge
and Development

'Targeting his analytical skills upon interactions among development professionals and their in-country interlocutors, Grammig encounters and empirically unpacks the cultural constraints to our notion of Technical Assistance. While others lace their development discourse with imprecise ideas about "culture", Grammig is laser-sharp in his theoretical stance and empirical framework. The results are sobering and provocative, with important implications for social science's contribution to the development process.'

William Partridge, Lead Anthropologist
Latin America and the Caribbean Region, World Bank

Development assistance often fails to improve managerial and technological capacity. The reform of these projects has been a widely acknowledged challenge for three decades.

Whereas interpretative social sciences have enhanced knowledge habits, learning and participation within commercial firms, in development assistance the politics of international relations has frequently rendered project implementation problematic, or even counter-productive. Moreover, technical knowledge is modified in substance and meaning when it passes from one social context to another.

This book seeks to overcome this legacy by theoretically and empirically demonstrating how development practitioners shape the organizational, social and intercultural dynamics of project implementation in industry.

The author demonstrates a fieldwork approach used by French contemporary anthropologists to grasp the cultural and social components of high- and low-technology projects as he considers and defines:

- the role of technical knowledge and know-how;
- project dynamics;
- constitutive sociocultural processes and communication structures;
- management tools for cultural distance.

Grammig argues for a limitation of planning and management in development agencies and proposes tools definable by the practitioners concerned for each context.

This work will be of essential interest to development practitioners, planners and evaluators, and to students of innovation, industrial development, international technology co-operation and project management, science and technology studies, technical change and intercultural communication.

Thomas Grammig, after studying chemical engineering in Frankfurt, spent 6 years as a consultant in Technical Assistance. Following studies in development economics at Sussex University, he obtained a PhD in anthropology at the EHESS in Paris. Since 1996, he has served as Academic Director in the *Leadership for Environment and Development* (Lead-Europe) foundation in Geneva.

Routledge studies in development and society

Searching for Security
Women's responses to economic transformations
Edited by Isa Baud and Ines Smyth

The Life Region
The social and cultural ecology of sustainable development
Edited by Per Råberg

Dams as Aid
Anne Usher

Politics of Development Cooperation
NGOs, gender and partnership in Kenya
Lisa Aubrey

Psychology of Aid
A motivational perspective
Stuart Carr, Eilish McAuliffe and Malcolm MacLachlan

Gender, Ethnicity and Place
Women and identity in Guyana
Linda Peake and D. Alissa Trotz

Housing and Finance in Developing Countries
Edited by Kavitta Datta and Gareth Jones

Peasants and Religion
A socioeconomic study of Dios Olivorio and the Palma Sola
religion in the Dominican Republic
Jan Lundius and Mats Lundahl

Environment and Development in the Straits of Malacca
Mark Cleary and Goh Kim Chuan

**Ethnic Minorities, Indigenous Peoples and the Impact
of Development**
Edited by Gerard Clarke and Alan Rew

Technical Knowledge and Development
Observing aid projects and processes
Thomas Grammig

Technical Knowledge and Development

Observing aid projects and processes

Thomas Grammig

London and New York

First published 2002
by Routledge
11 New Fetter Lane, London EC4P 4EE

Simultaneously published in the USA and Canada
by Routledge
29 West 35th Street, New York, NY 10001

Transferred to Digital Printing 2003

Routledge is an imprint of the Taylor & Francis Group

Typeset in Baskerville by
Prepress Projects Ltd, Perth, Scotland
Printed and bound in Great Britain by
Intype London Ltd

British Library Cataloguing in Publication Data
A catalogue record for this book is available
from the British Library

Library of Congress Cataloging in Publication Data
Grammig, Thomas, 1960–
  Technical knowledge and development: observing aid projects
  and processes / Thomas Grammig.
    p. cm.
  Includes bibliographical references and index.
    1. Technical assistance – Developing countries. 2. Technology
  transfer – Developing countries. I. Title.

  HC60.G644 2001
  338.9'26'091724–dc21   2001019765

ISBN 0-415-25364-0

Contents

Preface

The institutional (aid politics) and organizational (donor inertia) obstacles to producing anthropology of development efforts are immense. I could use the material resources and, more importantly, the motivation from 6 years as an engineering consultant mostly in energy projects funded by the US Agency for International Development (USAID) and the World Bank. Pierre Achard, Allan Hoben, Mick Howes and Michael Schönhuth encouraged me to try. Gérard Althabe and Monique Selim have taught me ethnography and made this research viable. Christine Jones translated and improved my writing. I would also not have been able to get this far without the generous advice of Raymond Apthorpe, Michael J. Chadwick, Jean-Pierre Chauveau, Kim Forss, Georgia Kaufmann, Arturo Lara Rivero, Anne LeNaelou, Fabrizio Sabelli and Michel Tibon-Cornillot. Finally, I owe the opportunity to produce this text to those who appear anonymously in these pages and I am deeply grateful for their trust.

Some management fads spreading in business schools actually damage commercial companies and are later uncritically applied to development efforts. Knowledge management could be a new one. In a global market where most power rests in the control over science and technology, 'managing' technical knowledge can be a versatile smokescreen. If I have contributed to filling rhetoric with ethnographic evidence of knowledge exchanges, reading this will be worth your time. Technology can enhance a developee's autonomy, affirm a developer's identity and even alter the forms that globalization takes. Little of this occurs because most developer–developee efforts disfigure their skills instead of using them for the interests of both. When developers and developees mutually admit this, they radically change their relationship.

Often, engineers discover that their skills are impregnated with social and political ambitions, and then they become social scientists to undo what their education made them believe and want. I am quite typical in that respect. In the North–South context, ethnologists are much better equipped for that discovery because their discipline came about within it. They do not need years of practice to face their assumptions about progress and social history. However, the ethnologists' problem of reflexivity is not made smaller because participant observation only produces insight into social reality when it is a shared effort of observer and observed. I certainly savour bringing my understanding to interact with the understanding of the protagonists of development efforts. Many of those who appear in these pages had similar feelings when we talked about the complexity of the development effort. I have taken a few prudent steps further into social theory and technology studies and assert that these steps are real opportunities for development assistance and for technology co-operation in industry.

Thomas Grammig
email: trgram@compuserve.com

1   Introduction

Beyond the common approach of examining development aid for its impact, we investigate its potential. In the area of technical assistance in particular, there is ample scope to explore the microlevel[1] and use theory never considered before to demonstrate that potential. Rather than only criticizing what is going on in technical assistance, we explore how it could evolve.

Aid practitioners are submerged in project implementation problems, reacting passively while searching for explanations. The insider perspective of technical assistance remains far apart from the outsider perspective, where planners and the public use vague hypotheses to form opinions. This study renders the two perspectives mutually intelligible. Based on the idiosyncrasies of each situation, a theory of practice explains the practitioners' daily efforts to understand each other and the resulting dynamics of an encounter between developers and developees. Throughout their careers, the developers and developees toil within the cultural dimensions of the economics, the technologies and the management on which their careers depend. Their dilemma is how to transfer technologies between societies. Without normative reactions, this research approach follows their intentions. Our premise is that modifying practitioners' work conditions will have far-reaching consequences for management, evaluation and social science research about technical assistance.

1.1 Historical context

The term technology transfer appeared in the 1950s in post-war reconstruction plans. Many development agencies now use the broader terms of *technical assistance, technical co-operation* or *building technological capacity*. In general, these differ only stylistically, and technology transfer remains the most persistent conceptual blank spot in all development aid by the industrialized countries. Technology transfer still appears in most policy packages produced for developing countries. This is understandable since changes in technology were the prime factor behind industrial expansion in the twentieth century, and are possibly responsible for 80 per cent of gross domestic product (GDP) growth in the North.[2] Actual technology transfer is accelerated by the economic logic of globalization. Nevertheless, little is known about how to influence, alter or orient such transfers.[3] Only a handful of writers, such as Denis Goulet (1989), have taken the time to explain what we do not understand about technology before calling for greater control and a harnessing of technology transfer to developing countries.

There are approximately 370,000 consultants, experts, advisors and volunteers engaged in technical assistance programmes world wide (Fry and Thurber 1989: 4).[4] There is constant experimentation with project planning, the type of technology, the origin and training of practitioners and the terms of their engagements. In 1960, four-fifths of the developers were English or French nationals, whereas today those countries account for only 8 per cent (ibid.). Japan, the USA, China, the Netherlands, Italy and Australia are similarly represented. Over time, there has also been a movement away from hardware to software transfers, i.e. from embodied to disembodied technology. The proverb 'Teach someone how to fish and he can feed himself' perhaps best summarizes the strategy behind that shift. Furthermore, the length of the engagements has changed significantly. While a few make careers in technical assistance, many others seek engagements for periods ranging from a few weeks to months. It often seems that every type of modification possible has been attempted, however little consensus has emerged on their respective merits.

Over the last two decades, social scientists have pursued new approaches to studying technology[5] and intercultural relations. In both fields of research, there is an urgent need for conceptual innovation. The combination of the two fields, that is to say technological changes because of intercultural relations, remains futile to the point where even the scope of the issue is not clear. For example, within the private sector, the first car factory in the USA with Japanese technology (GM–Toyota joint venture) incurred losses and turned out faulty cars for the first 6 months of its operation.[6] The plant achieved a reasonable level of efficiency only when most US staff in lower management were replaced by Japanese staff. Cars are undoubtedly American artefacts and Americans are accomplished car manufacturers. Why did Japanese technology require Japanese managers? Was it because of the assembly lines (the embodied technology) or the Toyota management style (the disembodied technology) or because of something outside the plant (general USA–Japanese relations)? It has been argued inconclusively which issue could explain the problems in operating the complex plant. The internationalization of business has created more research than development assistance. However, the simpler economic logic of business does not seem to lead to more effective research. Firms continue to seek competitive advantages in globalized markets through technology co-operations, and 'cutural problems' are blamed when most mergers and joint ventures fail.

As part of development aid, the US Peace Corps and the Canadian International Development Agency (CIDA) made significant efforts in the 1970s to evaluate technical assistance and improve project planning and the preparation of developers.[7] However, the results obtained have not allowed us to predict, let alone alter, the performance of Peace Corps volunteers or CIDA experts. Other agencies, such as France's *Ministère de la Coopération*, maintained an evaluation department for only 1 year (1986), finally concluding that their development aid did not require evaluation (Freud 1988). For a period, it seemed that non-governmental organizations (NGOs), which offered decentralized institutional support, might provide a solution. Governmental agencies appeared to be less effective than the NGO model of direct collaboration between local institutions. This model, however, has not flourished as governments have maintained a central role in economic growth. In some sectors of development aid, NGOs have appeared and they thrive, whereas in others, and especially in technical assistance, NGOs remain weak. Compared with bilateral aid, multilateral programmes have increased and subsequently decreased in importance over the past decades.

Analysing technology and intercultural relations in combination is an obstacle to filling the conceptual blank spot of 'technology transfer'. Despite constant changes between 1960 and 2000, most attempted improvements have failed. In fact, on the contrary, the changes have brought and continue to bring diminishing results. While we now know more about *intermediate technology*, about *appropriate technology* and about industrial policy in successful technology-importing countries such as Japan, South Korea or Taiwan, for the majority of the 160 developing countries technical assistance remains an uncertain endeavour.

By submitting two projects as different as Appui Technique and Autogeneración to the same critical analysis, we uncover common mechanisms at work. Like many other institutional practices concerned with modernizing society, technical assistance is largely misunderstood and obscure. Thus, the first step in research, the definition of an object of study, is the most difficult and is given most attention. Can we speak of technical assistance as an objective enterprise despite the diversity of agencies and the plurality of discourses on development? Is there something in its enactment, in its style and form, that gives it a coherence beyond the goals attributed by the official jargon, i.e. aid, assistance and development? Elements common to several diverse projects, this study suggests, allow us to construct a viable object of study that is presently called *technology transfer*. The ultimate aim is to provide practitioners and researchers with a coherent theoretical model to understand technical assistance, moving from the particular to the general.

1.2 Development events observed

The following remarks introduce the context of the two case studies but they are not part of them. Because of the nature of the analysis to be undertaken, these circumstances do not reduce the significance of the case studies to a critique of technical assistance more generally. The particular sectors and organizations are not of significant interest so much in themselves, but rather as the events described are representative of innovative efforts undertaken by reputed specialists from the high-technology and the low-technology ends of technical assistance. Most conventional technical assistance is located in between these two ends. The case studies are therefore representative of the institutional and the technological dimensions of technical assistance but not of other dimensions.

1.2.1   Case study 1: Appui Technique[8]

Chad is one of the poorest countries in the world, situated just south of the Sahara desert. In 1989, after 15 years of civil war, the Chadian government had little institutional capacity outside the military and the police. The International Bank for Reconstruction and Development (IBRD or 'World Bank'[9]) designed a typical structural adjustment programme (SAP) to give this government a financial base. Given the political instability, this SAP was accompanied by a soft loan of $US14 million to ease the immediate social impact.[10] The IBRD sought advice on enabling the informal sector of the Chadian economy to supply new products, thereby creating employment opportunities and replacing imports. Other measures such as infrastructure investments and small-scale banking support were also chosen by the IBRD.

After comparing proposals from leading NGOs in different countries, the IBRD signed a contract with the *Groupe de Recherche et d'Echanges Technologiques* (GRET) in Paris to implement the project 'Appui Technique'. The estimated budget was about $US1 million. GRET had a history of working closely with the official French aid agencies, following the 1981 election victory of the Socialist Party in France. Appui Technique was their first contract with the IBRD and was a chance to demonstrate their competence. The Chadian context presented an opportunity for a deliberately modest start in industrial policy. GRET assembled a team of Chadian and French experts to train artisans in the capital city, N'Djaména. Chadian artisans had failed in the past to compete with imported machinery. The experts invited the artisans to learn through welding prototypes of oxcarts, grain mills and other machinery with the tools and the raw materials available locally. The Chadian experts took this opportunity to enter *terra incognita*. The Chadian *Ministère de la Fonction Publique et du Travail* at least wanted to follow and understand what GRET's experts were undertaking. But beyond participating in the paperwork, the Ministère did not interfere with Appui Technique.

Appui Technique was concluded in 1995. According to GRET's evaluation report, the Chadian artisans had sold machinery worth $US140,000. GRET's report provided a constructive assessment of Appui Technique. However, the IBRD never responded to the report, nor has it sought GRET's advice since. The French and Chadian experts currently work in other fields of technical assistance, while the Chadian artisans continue struggling to make ends meet. As a part of Chad's debt to the IBRD has since been written off, one cannot conclude that Appui Technique was a net loss to the country.

1.2.2 Case study 2: Autogeneración[11]

From the subsistence context of Chad, we move to the industrialized context of Mexico, where the engineering education and industrial corporations are similar to those in the USA, to compare the transfer of manufacturing knowledge with another similar objective in electric power generation. In 1990, the Mexican Energy Agency attempted to pursue President Salinas de Gotari's 'free market' industrial policy guiding private sector investments, instead of the classic Mexican government's policy of command and control. Cogeneration technology was deemed appropriate to demonstrate this shift. Indeed, other governments successfully implemented energy policies around cogeneration.[12] Such a policy shift required sound knowledge of the economic parameters for energy investments. Thus, the Agency sought advice from a renowned consulting company specializing in cogeneration, Hagler, Bailly, Inc.

The IBRD has given financial assistance to the Mexican national utility company since the 1950s. This is the type of client that this bank was set up to deal with. Always keen to include technical assistance in a loan package, the bank suggested using $US600,000 for cogeneration feasibility studies through the Energy Agency out of a $US450 million loan package for the Mexican utility company. Hagler, Bailly, Inc. conceded favourable terms in the competitive bidding for the project 'Autogeneración', seeing this as an opportunity to produce 'a landmark study'. The work was an extension of what Hagler, Bailly, Inc. was doing in the USA, and no particular problems for cogeneration investments were anticipated in Mexico. To intensify the transfer of technology, the Mexican Energy Agency assembled a team of Mexican experts to work with the team from Hagler, Bailly, Inc. For 2 years, up to twenty-five engineers and economists compiled and analysed data from steel mills, chemical plants and other energy-intensive industries, transforming the data into decision criteria for private investors. Their reports recommended, for example, an investment of $US196 million in the plant XY, using the gas turbine ABC, operating at 57 per cent thermal efficiency and producing 2,000 GWh of electricity annually, and concluded an internal rate of return of 28 per cent to the investor.

By the end of 1993, Autogeneración concluded. However, the Agency has yet to pay all of those who worked for it, and none of the Mexican experts have continued to work in the field of cogeneration. Of the forty feasibility studies produced, only four investments were pursued, but not by companies involved in Autogeneración. As is often the case in technical assistance, it is difficult to determine the impact of Autogeneración. Neither the Agency, the IBRD nor Hagler, Bailly, Inc. attempted to define the *lessons learned*. Although transfer of technology is an evasive objective, difficult to grasp empirically let alone to evaluate, such projects are repeated with different configurations, in Mexico and in many other countries, and all seek to fulfil goals similar to those sought within Autogeneración.

1.3 Basic questions

Basic questions are those without a definitive answer, such as 'which are the appropriate research methods?', 'what are the roles of individual actors?'[13] and 'what influence do aid organizations have?' We can produce elements of answers and add to related research on these basic questions.

Drawing upon related research, the following section will begin to point to answers to such questions. Technical assistance is a field of research that presents itself in the form of 'projects' that employ 'experts' to realize a transfer of knowledge between countries. A simplistic approach would begin with the idea that such projects are simply autogenerative and self-sufficient events[14] with little substantial outcome. In such a simple analysis, jargon is offered by the experts, everything happens as it should 'on paper' and nothing changes. An observer witnesses yet another 'non-event'. The events of Autogeneración and Appui Technique, however, were spaces of communication where the reflections of the actors were much more extensive than the reflection that went into conceiving them. Rather than focusing on project objectives, then, we examine the life-world that these objectives presented to the actors. The unique dynamics, born of the events themselves, are beyond recognition or control by the development agencies. By foregrounding the circumstantial context of the encounter, we demonstrate how much the professional relations were defined by the everyday interaction. The actors did not proceed blindly because the projects themselves were ill defined. The principle source of their confusion was the overwhelming complexity of the encounter. Thus, despite the limits of the predefined objectives, these projects were far from being non-events. Much took place within these events – within the everyday interactions – that requires a sustained analysis.

This study describes how developers (foreign experts, consultants and volunteers) and developees (local experts and target population) struggle with power and cultural distance.[15] The contemporary relationships in technical assistance embed technical knowledge in power and cultural distance in specific ways. Even if foreign experts succeed in transmitting their knowledge, they cannot transfer the capacity to use this knowledge effectively to local experts, the beneficiaries of the formers' development ambitions. The theoretical modification of the role of the foreign expert according, for example, to a model of exchange based on a collaborative, rather than a colonial model, has not sufficed to produce effective assistance and local autonomy in actual development projects.

Is it possible to draw out the significance that aid and assistance assume between industrial nations and countries said to be developing? This is the principal focus, and thus the tenets of development theory do not constitute the subject of this research as such. The analysis is not technological, but rather offers a cultural analysis of the institutional and managerial practices prevailing in the 1990s. Power may seem to operate in a rather straightforward manner, as economic power, but cultural distance plays a complex and more pathetic role. The analysis will provide a diagnosis of the destructive potential that cultural distance holds within power, notably by understanding the failure of planners to receive feedback from the field, suggesting that this lacuna is an institutional expression of cultural distance perpetuated through development agencies. The failure of technical assistance, and the sheer impossibility of feedback, are often accepted and are attributed to the problems of 'less-developed' societies. Even those agencies that try hardest to obtain such feedback admit that they find it almost impossible.[16] Therefore, the object of this study derives from an idea that places assistance as the practical extension of the discourse on development. Thus, if technical assistance puts into practice a theory of development, its significance can be interpreted in very different ways. For some, development is a tool of Man's liberation from material constraints. For others, it is no more than a contemporary form of domination in today's globalized economy. It can also be perceived as a charitable undertaking that soothes the conscience of wealthy nations, or as the vehicle of the myth of modernity, or alternatively as a humanitarian gesture.

The agencies, firms and associations that offer technical assistance pride themselves on the consistency of their discourse on development, and the supposed efficacy of their aid and assistance. I will show to what extent, on the contrary, they lack both the enlightenment and means to put these methods into practice. I want to emphasize precisely the disparity between the discourse and the practice of development and to discern and deconstruct the logic of the exchanges between the actors in a project. Using two case studies, I demonstrate what this enterprise is really about in its practical, everyday functioning. It is only through the work of the individual actors that meaning and substance can be given to technical assistance. Thus, let us start with the questions 'how do they construct meaning during the course of their collaboration?' and 'where do they find the necessary social meaning for their collaboration to implement what is written into their tasks?'

1.4 Results

Three latent processes were responsible for the abyss between theory and practice in both projects studied. These originated in the actors' combined faculties and are inherent in all efforts to use technology in a foreign society. To endorse the definition of the latent processes, as the principal result, we will show how agencies' management and evaluation habits have come to reflect these processes. The processes are certainly unknown to planners and evaluators at present, but they account for these processes in an implicit manner. Tracking this reflection, we will subsequently see how agencies can help actors seize upon these processes. Finally, having understood the idiosyncrasy of each implementation, as well as the current limits to evaluation and management, we can propose implementation tools that will enable the actors (local artisans, foreign engineers and other experts) to seize upon their mutual learning.

Significantly, in both Autogeneración and Appui Technique, there were no conflicting interests that could explain why the local and foreign actors could not advance. Had all the actors adopted a more constructive attitude, Hagler, Bailly, Inc., GRET (the development agencies) and the local institutions involved would undoubtedly have obtained results that more closely approximated what they had initially envisioned. The institutional and personal interests played compatible, even complementary, roles. The local governments, the IBRD and other local partners could all have gained.[17] However, they lacked the management skills needed to understand better how to meet each other's needs and expectations, to see what they were achieving and how. Their achievements disappeared like water running through their hands. Further, the outcome cannot be attributed to any one guilty party, hidden agenda, or alienating power of technology at work. Rather, it needs to be understood in a much more general way, relating fundamentally to questions of cultural distance.

Although there were some successes, the actors were unable to build on them in their encounter. For example, an artisan participating in Appui Technique learned to produce a groundnut mill and built one within a few days in his own workshop.[18] The experts in Appui Technique could not have imagined such effective technology transfer. And this artisan could not imagine telling the foreign expert about his learning. Neither side could see the results of their efforts. Only the actors can discover their strengths and weaknesses to overcome such distrust. The apparent tasks at which the actors failed – quality control by artisans and approximations for feasibility studies – would have been insignificant if the local and foreign actors had recognized what really separated them.

The actors' means of pursuing their interests were limited by the tacit rules and professional biases in the companies where they began their careers, their prior achievements and failures in technical assistance and, most importantly, by the original dynamics produced by their exchange of knowledge. These were different reasons for the unique nature and the autonomy of project implementation from the general context.[19] The context had been taken into consideration during the planning between the development agency and the recipient institution. However, the actors' use of technical knowledge led to events that no planner could foresee.[20] We will establish to what extent this has to be the case.

We submit the two projects to the class of human endeavours that Peter Berger has called 'Pyramids of sacrifice',[21] i.e. calamities programmed with the best intentions. Technical assistance pyramids are created in the abyss between theory and practice. Their origins resemble the origin of those described by Berger, the void between political ethics and social change. 'What stuns and paralyses the mind is the use of realistic means in the service of a metaphysic both rigorously rational and delirious, the insensate offering up of lives to a petrified concept' (Berger 1974: 8). Writing anthropology about a modern phenomenon, such as technical assistance, reverses the hierarchy of planners and actors – the latter determine what can be pursued. Power and cultural distance always shape technical assistance, but pyramids of sacrifice succumb or soften when latent processes become visible to their actors.

Considering developer–developee relations, one might be inclined to write a management manual on how to avoid pyramids. Currently, there is a dearth of management literature and the manuals are generally ignored. However, there are very good reasons for this. A manual can only refer to a general context, whereas the dynamics that evolve during implementation cannot be anticipated and thus cannot be written in a 'how to' format.[22] Rather than a manual, we will describe tools that actors can use, distinguishing the tools for exo-social and endo-social situations.

Because of the very nature of the developer–developee encounter, it is not necessary to move beyond the two cases analysed here. While the particular dynamics of any project are unique, we aim at the actors' understanding of an event, as this alters that event *at the same time*. One general condition will be obvious: developers and developees always co-author project implementation. Their encounter will change when both sides comprehend what is going on with new perspectives. Our results reverberate the actors' efforts to attain the social and political objectives their technical knowledge is meant to reach. In commercial technology co-operation the encounter is less ambitious, but its nature similarly problematic.

Further, demonstrating how to observe these developer–developee encounters is, at the same time, pointing towards different encounters. Writing from my perspective as an observer, I anticipate how the practice can change. Ultimately, this happens only from within the institutions that employ these actors. There is no need to postulate any particular change. Current management and evaluations are examined only to illustrate the latent processes, and indeed to advocate the reconstitution of the processes elsewhere. Helping actors to understand what is going on around them, improving planners' knowledge of the limits of their planning and augmenting actors' confidence to document practice and to force feedback into development agencies are all closely connected. All this starts with the actors' capacity to observe.

1.5 About the text

Referring to my participation in the events, I use the first person and past tense to describe what I did. This text, written after years of research, uses the third person ('we' as subject). The present tense refers to the arguments the author suggests to the reader. Observing first hand the daily functioning of two technical assistance projects was only possible by first being hired as an engineering expert, and then by later requesting permission to study the events through taped interviews and meetings. Permission, it should be noted, was then reluctantly granted. As with most of what follows, the need to circumvent this particular barrier to research is a small manifestation of 'big picture' concerns in development agencies. This text is thus the effort of a practitioner trying to understand the mechanics of technical assistance. Its motivation was experiencing the frustration of 7 years of technical assistance ineffectiveness. Now, with 8 years of work split between engineering and anthropology, I find opportunities for applications despite the institutional obstacles for feedback from technical assistance practice that exist within development agencies.

The text should be read as one crosses a canyon. We start by looking from afar, from the high ground of academic debate. During the steep descent, Chapter 3 locates the signposts for orientation before we arrive down at the bottom in a wonderland described in Chapter 4. Getting back up always takes longer. The first latent process concerns what is flowing through this creek in the middle. More theory is introduced to provide lampposts and oversight. The diversity of these enlarges the horizon. Finally, in section 5.5 we reconsider why it is so difficult to look at the creek from the top. To consolidate the exercise, we draw a map in Chapter 6.

The empirical basis of the events appears only to help the reader understand how I produced this account. As in the famous story of the blind men who touched an elephant,[23] I *talked* to those whom I observed about their impressions. By explaining these exchanges, the reader can follow my anthropological account of technical assistance. Specialists, developers and developees may find this account familiar and may liken it to or contrast it with their own experience. Other readers, unfamiliar with technical assistance, may find this account rather peculiar, and even surprising. The results of the case studies are included in detail also to give an inside perspective of the events. This should assist the reader with the theory and with the examples used subsequently to describe the latent processes. However, the latent processes can be understood only with the examples used there, without referring to the case studies. Individual actors are briefly introduced as they appear in Chapters 3 and 4.[24]

Originally, I had to assemble each case study (see Chapter 4) before isolating the interfaces and the project dynamics. You, the reader, might begin there, but, unless you are going to participate in similar events, you will find the results of the case studies very circumstantial. Nevertheless, there are many lessons to be learned about technical assistance, and one has to struggle with these in order to produce such a case study. These lessons are announced first in published research (Chapter 2) and then in the research strategy (Chapter 3, which begins with a comparative table of the cases). The case studies are also important for a reader who never had a chance to visit 'a project'. Practitioners might read the case studies (Chapters 3 and 4) rapidly and move straight to the processes (Chapter 5) if their interest is mainly planning and evaluation, or to management (Chapter 6) if they are engaged in implementation. Addressing practitioner and research interests together, the logic of project implementation is defined in its limits and its determinants. More anthropologically inclined readers, academics or consultants will link the processes back to the research strategy. A reader might go back to the case studies also to reconsider under which conditions these processes were upheld by the actors through technology and intercultural communication.

The revisiting of some concepts should help a reader to reproduce such an account of technical assistance events. The cases can also be useful to the reader who wants to interpret development anthropology differently. Later on, the evidence for the latent processes can be related to different interpretations of data (the construction of the intelligibility). Some accounts serve to criticize and effectively dismiss technical assistance, and other accounts argue for continuing technical assistance. The goal of this particular account is to improve it. The general assessment of development anthropology and ethnographic experiments are pursued elsewhere and this is a contribution. Nonetheless, our interest is in technical assistance practice as such. Between theory and empirical description, much insight is gained. The definition of the latent processes can be applied, consolidated, abandoned and improved. They point to different bodies of literature and provide a basis for further comparisons. Extrapolating from the events, the proposals for managing project implementation are meant to inspire further inquiry into these subjects.

2 Development anthropology

The first school of thought within the social sciences to assess the validity of economics and engineering in development aid was development anthropology in the USA. Understanding the past experience of social science research in development aid points to the particular interest of this book. There is no scientific agenda in this analysis of Appui Technique and Autogeneración, but research is also a social practice (engaged in by those who pursue it) and it ought to be questioned with respect to its context. Several recent publications arrive at conclusions partially reflecting the latent processes.[1] Despite the focus on practice within this book, the results emphasize the limits of economics and engineering.

The following sections point to the conclusions drawn by development anthropologists about their relationships with development agencies and highlight some of the results that they achieved in the bilateral aid programmes of the US Agency for International Development (USAID) in the 1970s and 1980s. Their comments point to important conditions regarding their work, however the appropriate scholarly account is beyond the present scope. Thus, my remarks are not well substantiated, but, taken together, an appreciation of the issues will emerge. Introducing the debate between development anthropology and an anthropology of development prepares the context for the case studies to be examined. We emphasize the following aspects of development anthropology specifically:

- high-technology and low-technology in industrial contexts;
- development practice as the source of the objects of study, the quality of the fieldwork as central rather than the reconstruction of the observed;
- actors, foreigners and the locals considered as active and passive mediators, and the researcher is only a passive mediator;
- while implementation is only a part of the 'project cycle', the cycle should not orient the analysis and, most importantly, the actors themselves need to be understood as transcending the project cycle;
- the project is a microsocial space of communication with its own logic.

2.1 Phases in development anthropology

Social scientists have been working in development assistance since the independence of developing countries. Throughout the first two development decades, these experiences were isolated and few general conclusions were drawn.[2] One of the first attempts to remedy this was made by Allan Hoben, who reviewed and consolidated what had been documented before 1982. He found that, because of the differences among the pioneers, the potential roles and the involvement of anthropologists were diverse. They were alternatively mediators, agenda clarifiers, advocates, cultural brokers, trouble-shooters, interpreters, go-betweens, etc. Hoben concluded that 'no coherent or distinctive body of theory, concepts and methods' existed (Hoben 1982: 349). This situation continues today. However, what appears clear is that there are four important conditions for every involvement:

Condition 1:   the development agency or donor.

Condition 2:   the particular field addressed.

Condition 3:   the social context.

Condition 4:   the anthropologist as an individual.

Development anthropology has accumulated a considerable amount of empirical data documenting development practice – but no clear trends towards 'conceptual closure' appear. Taking Roger Bastide (1971) and Glyn Cochrane (1971) as the pioneers, the subdiscipline has been emerging for 30 years now, which raises doubts about its feasibility, considering its lack of advancement.

   The bulk of development anthropology was produced for one development agency,[3] the USAID (condition 1). The focus continues to concern agriculture and health (condition 2) and predominantly rural areas in the least-developed countries (condition 3). Understanding these three conditions permits a defining of the scientific objects and the potential of development anthropology. It has also been pointed out repeatedly that the social context, rural populations and the fields addressed – agriculture and health – reflect the historical conditions in which anthropological research appeared.

   Nevertheless, there seems to be no reason why these conditions remain primary. The powerful results from anthropological research in industrialized countries indicate that the colonial context had little influence on the anthropological methods and tools (heuristic and epistemological). The 1990s have indeed seen an explosion of anthropological research, particularly in the USA, the UK and France. What keeps anthropologists from addressing in the South the same fields in which they invest in the North? Can these lacunae be traced to the colonial heritage of the discipline? What administrators and other decision-makers in the development agencies understand by development anthropology represents the greatest obstacle for this scientific subdiscipline to emerge. Their prejudices haunt anthropologically inspired qualitative social science and continue to give primacy to the above-identified conditions 1, 2 and 3. Sometimes, very specific contributions from anthropology are sought; shortly afterwards, these are no longer of interest. Possibly, the succession of policies in development agencies is too fast for development anthropology to keep up with and to pass through the different phases of consolidation required of an academic discipline.

Today, the need to consolidate singular applications remains an urgent task. Many publications continue Hoben's work,[4] and most still reflect the three conditions identified above. In development practice, anthropological research is possible in rural areas, certain fields and certain development agencies. The critical mass of research results required to go beyond these conditions and most importantly beyond the prejudices established has probably not been reached. According to those authors who continue Hoben's work, the innovative role of USAID is primarily the result of the 'New Directions' period (following legislation of the 1973 United States Foreign Assistance Act).[5] The results of this period have been Country Development Strategy Statements (CDSS) and Social Soundness Analysis (SSA), which contain bits and pieces of social analysis on co-operatives, irrigation, crops, farming systems, livestock, dams, resettlement and migration, and health, especially health care and family planning. In these cases, however, social science was simply added on to the current practices of project identification, planning and evaluation.

The most comprehensive source book is John van Willigen's (1991) compilation of the 530 major applications of anthropology. Among the 530 applications, twenty-nine concern agriculture in developing countries and only four in industrialized countries, whereas one concerns the law in developing countries and seven in industrialized ones.[6] It should be noted at this phase of an emerging subdiscipline that, if particularly successful applications were to be gathered, it would certainly leave blank spots in the spectrum of applications. The exploratory spirit, which defers the tasks of synthesizing, is reflected in the anticipation of obstacles to come. The results of development anthropology need to be explained before they became unfashionable:

> Despite the need for the kind of advice provided by social analysts with extensive country knowledge, the concepts and terminology of social analysis developed in the late 1970s in AID are currently fading from the development vocabulary, and they are not likely to reappear in the same form. ... In time, perhaps the label 'rural investment advisor' will come to mean social analyst.
>
> (E. Greeley in Green 1986: 245)

By 1986, the 'accelerated development' policies in the IBRD and USAID had moved development practice away from 'basic needs-oriented' programmes (central to 'New Directions') towards policy dialogue, private sector involvement, institutional development and (being constantly rediscovered) transfer of technology. The social science lessons learned during 'New Directions' were not assimilated. It appeared that the objects of research could not be authenticated independent of institutional interests. Much experience has been lost because of these problems, and especially the contribution of development anthropology to the implementation of projects in urban contexts (Mason in Green 1986: 141–59).

The New Directions period was, despite its limitations, an expansionist phase. This phase has ended; however, those researchers who published their results at this time pursue their engagements in development aid today. During the expansion, these researchers gained professional credibility. The credibility attained during this period remains attached to their names, rather than to the results as such (condition 4).[7] How do development anthropologists describe their entry conditions to a development agency? William Partridge identified early on that 'the challenge of being an effective anthropologist is met only by studying up[8] the organizational hierarchy in which the project is created, shaped and maintained or abandoned' (Partridge 1984: 3). Erve Chambers observed in his introduction to the volume edited by Wulff that the most exciting aspect of 'final analysis' was the fact that the case studies were prepared by enthusiastic people just entering the productive stages of their careers (Chambers in Wulff 1987: viii). All this suggests that, during this phase, the opportunity to study up opened and closed again. Those few who succeeded in studying up, i.e. who first gained insight into the organizing principles of their employers, and then also succeeded in applying this knowledge to their work on subsequent projects, are currently able to continue applying their results. Other anthropologists cannot do likewise because they did not begin during the 'New Directions' era, nor do all facets of their professional background match those who can continue to apply their results.[9] Interestingly, many anthropologists who undertake development work today use different labels (Little 1992; Partridge 1994; Curry 1996). What was called 'development anthropology' becomes 'irrigation studies' or 'livestock research', while applying the same methods of analysis.[10] Without exploring this phenomenon in depth, one can translate this conclusion into a broader hypothesis about the relationship between applied research and development agencies.

Possibly other fields and other social contexts hold more and different results for development anthropology. However, it is impossible to demonstrate this because the opportunities for research are not there, owing to the lack of funding, and because of questions of access to the development practice. Access is key because participant observation is the primary methodological tool used. Social scientists must justify to the agencies why they want to observe the implementation of development projects.[11] Without potential results to propose, there is little opportunity to find out what is going on in development practice, in projects, in the administrations or during negotiations. Thus, the inroads gained by anthropologists during the expansionist phase have not led to an expansion of the initial experiences.

2.2 Specificity of the phases

To verify this conclusion, the specificity of the 'New Directions' phase must be explored with regard to the development agencies, the field and the social context (the first three conditions mentioned above). Thus considered, the results obtained during 'New Directions' can:

- represent a singular experiment, subsequently abandoned;
- remain isolated without lessons applicable to other donors, other fields or other social contexts;
- constitute a basis for new approaches to the same objects.

Frequently, development anthropologists have identified some aspects of their engagement with respect to the development agency. 'The inclusion of anthropologists on project teams currently is compatible with the rhetoric of donor agencies' (Robins in Green 1986: 17). 'The skilful manipulation of conflicting, or at minimum, different interests is difficult ... the short-term assignment asks of the anthropologist this mediation, but does not afford him/her the time needed to make the role credible' (ibid.: 68). If Partridge is right to identify studying up as the key to an effective engagement, it is logically coherent that these writers point to the agency's conditions for their work. If they could not do so, they would not be employed. However, are there specific conditions for working in USAID activities that can be identified? There are no such suggestions in the literature. Furthermore, some development agencies seek to learn from USAID, and as they have not been able to achieve much[12] this suggests that agency conditions are not the dominant ones.

The other two conditions identified concerning the field and the social context are rarely scrutinized in the literature. In agriculture, 'local level research will naturally lead to a critical examination of the appropriateness of technologies (and policies) offered by development agencies' (Fujisaka in Green 1986: 180).[13] In the fourth development decade, 'participation' has become the mantra[14] of development agencies (together with sustainability). But when providing advice on participation, no 'natural' phenomena surface, and with a less defined role advice is correspondingly weaker. Researchers have become increasingly tied to the specificity of the field of application. In health and in agriculture, anthropologists tend to work on understanding local knowledge systems. But contrary to agriculture, the anthropologist has less chance of defending the differences between local and Western practices in health projects. In health projects, anthropologists tend to concentrate on health-care administration:[15] 'Anthropology is conceived of as a discipline which helps raise "compliance" to a predetermined treatment regimen. Critics would argue that the anthropologist is used, not anthropology' (Kendall in van Willigen 1989: 300). Medical knowledge seems more autarchic than the agronomist's knowledge. Only in nutrition-related interventions can more behavioural factors become better accepted.

Both in health and in agriculture, local knowledge systems are centuries old and this gives the relativist[16] explanatory contribution of development anthropology a clear niche. Similarly, sustainable development, the other mantra in development agencies in the 1990s, is also favourable in that respect. Local knowledge about the biosphere is well encoded in society, and subjects such as indigenous resource management find an increasing audience (Chambers 1997: 26–29). Where local knowledge is less culturally and socially encoded, such a niche is more difficult to establish. This does not preclude an anthropological contribution regarding local knowledge in other fields because, a priori, any bit of social reality can be studied, but it gives a partial explanation of why such contributions have been less available. With respect to the field of developmental interventions, development anthropology seems constrained through changing vogues. Since anthropological results are at least partly specific to a field, the relative attention that development agencies provide to a given field can reduce or increase the contribution of development anthropology. The field is thus an indirect condition for development anthropology. It appears that this indirect condition, imposed via the field of intervention in vogue, is a more important determinant than a particular policy of the development agency (in this case, USAID) towards development anthropology as such.

This leaves condition 3, the social context of the developmental efforts, to be examined. The greater the power differential between the developer and the developee, the more effective development anthropology appears. Particularly in Wulff (1987), the most salient case studies with respect to the anthropological contribution[17] involve rural labourers and federal governments, tribal people and the US army, etc. Where powerful developers are radically foreign to the developees, an anthropologist is in control by virtue of his/her comprehension of the differences. The mediating role is enhanced by, for example, the coding of cultural differences in sophisticated questionnaires. 'Parker, based on her daily interaction with the people of the villages, made sure that King understood their needs with precision. King, understanding the workings of the government and accepted (marginally) as an insider, tried to cast these concerns in terms the government could understand and to negotiate about them on behalf of the villagers' (Parker and King, in Wulff 1987: 164). The responsibility and the potential impact of development anthropology appear essential for such research. Consider, for instance, Edward Green's reflection on his engagement:

I suspect that critics of 'establishment-approach' aid are fundamentally correct. However in the short term, my lifetime for example, I'm not so sure exactly how or even if power relationships can be fundamentally restructured. Yet even with all these uncertainties I feel anthropologists can and should participate in projects directly concerned with life-protecting and life-enhancing measures, while at the same time seeking ways to improve the condition of the poor in ways that are more structurally fundamental. For me, the Rural Water-Borne Disease Control Project has served as a vehicle for the realisation of some personally held humanitarian aims, while at the same time providing opportunity for professional growth.

(Green 1986: 120)

An anthropologist's normative stance is here at the very core of the engagement.[18] More than in other disciplines, the researcher chooses the objects of analysis with respect to a professional deontology. While this reflects the power differential in the social context concerned, it is certainly not causal, as the development anthropologists' choice of an object of research will not create control of the power differential between, say, a minister and the villagers. But insofar as the power differential is also expressed and encoded in social and cultural differences, development anthropologists gain influence to the extent that they understand how these social and cultural expressions of power function.

Therefore, the professional deontology will create insurmountable obstacles in contexts where an anthropologist represents a formerly hegemonic country. France is such a country, having a strong anthropological research tradition. Thus, in French research, early warnings in the 1970s have contributed to deflecting the inroads made by the USA. The following three prominent French authors have been influential by pointing to the political conditions of development assistance as well as by exploring research opportunities themselves:

The relations between developed nations and developing nations are called 'development assistance', in the best case. A whole population of 'experts' appears. Nobody has yet undertaken the essential task of a sociology of that assistance or of the expert. But, are we able to understand the conditions under which our [French] assistance is organized? We learn only by accident, due to the international scandals.

(Berque 1965: 433)

The only possible sociological object of analysis from a 'development project' is the project itself, its modalities, the complex formed by the developer and the developee, ... how it is planned and implemented, how it is perceived by those who are the intended beneficiaries and its objects (in the sense that their habits, their techniques, their mentalities are changed). No serious sociological analysis càn predate [in the sense of existing independent of] such a development project.

(Augé 1972: 208)

In development the populist ideology is institutionalised. This populism has succeeded in selling a number of products in the development market. Schumacher and Freire are the pioneers and emblematic figures. This populism is continuously reinvented ... the conjuncture populism/ anthropology/development is already in place in Cernea, Pitt and Hoben.

(Olivier de Sardan 1990: 479)[19]

At least at the level of publication, such warnings have been heeded in France more than in the USA. There is a tacit consensus among anthropologists in France not to engage in research for the development agencies (Amselle 1991).[20] The historic context of French *coopération* (official development assistance) imposes particular conditions that explain the reluctance of French social scientists. French colonialism was more assimilatory than the British version (Amselle 1990), and French *coopération* is less open to anthropological insight than the British DFID (Department for International Development). However, the reluctance among anthropologists is the decisive factor.[21] The assimilatory character of the colonization complicates the articulation of a professional deontology. Jean-Loup Amselle does not propose a further analysis of this tacit consensus concerning *coopération*, which relates to the general mould of the French *chercheur* in connection with the French state or rather with society. Professional deontology could therefore be one reason why development anthropology has not been thriving in Europe.

Unfortunately, such a comparison of development anthropology between different countries is difficult to operate. To understand whether it is the professional reluctance or the difficult operationalization of anthropological contributions which is limiting research, one has to know, for example, how many requests for proposals from development agencies created less than state-of-the-art offers.[22] There is limited anecdotal evidence of anthropologists refusing to respond and I cannot judge how typical that evidence is. Nonetheless, and without exaggeration, one can state that an anthropologist's approach to the power differential between development agencies and the developees is the core question behind development anthropology's fate.

In sum, there are two specificities to be stressed: the fields of intervention and the power differential between development agencies and the developees. When an expanding scientific practice ignores the founding principles, its advances can be limited. I am not aware of a publication addressing these two specificities. The lack of analysis of the driving forces of development anthropology contributes to the vivid reactions once these are challenged. The pioneers' reflections on their individual experience (see Green's quote above) should become part and parcel of the subdiscipline. The difficulty of establishing for whom anthropologists speak takes precedence over the choice of the object. Hoben's (1982) verdict that there is no coherent body of theory is still valid and, returning to the initial question on p. 16, although development anthropology has not yet been abandoned the results appear isolated and the debate about the definition of new approaches rages throughout anthropology.[23] Some pioneer development anthropologists, such as Michael Horowitz, maintain that they have not failed in the substance of their work but in the effectiveness of their communication. Others refute this by pointing to deficiencies of the results.

Before opposing these two positions with examples from Senegal and Haiti, the introduction of a potential alternative to development anthropology of the 'New Directions' phase will clarify the debate.

2.3 Actor-oriented approach in applied research

It took 10 years to digest and publish the inroads made during 'New Directions', roughly the 1980s. While the development practice has simply moved on to new modes, fads and paradigms, reducing the involvement of anthropologists, anthropological research increasingly studies itself, reducing the attention to its reception and application. There is no connection between these two phenomena,[24] between the changes in the political climate within development agencies and donors and the scientific changes in anthropology as a discipline.[25] However, those anthropologists who had already integrated their epistemological efforts on fieldwork situations into studying development practice automatically moved to the forefront in development anthropology.[26] This shift has been called for, and announced repeatedly, in indicating a move from development anthropology to an anthropology of development (among others, by Augé in the above-cited article[27] and by Bastide). During the 1990s, this shift was slowly consolidated in different fields and contexts of development practice.

Arguably, the beginning of a viable anthropology of development was found in *Encounters at the Interface: A Perspective on Social Discontinuities in Rural Development*, the culmination of 25 years of fieldwork in Latin America and Africa by Norman Long and his team (Long 1989). Among its many merits, Long's interface analysis ended 'the grand divide' in development anthropology, the epistemological charity towards 'less-developed societies'. Hitherto, it was not possible to say 'yes we'll study *also* the development agency (probably later on and with less attention than the intended beneficiaries)'. Long showed that to understand development practices requires examining both the developer and the developee in one and the same analysis.

An actor-oriented approach uncovers the interlocking intentionalities existing among those concerned in the development intervention.[28] Concepts such as life-world, agency (reach and horizon of an individual actor in social theory), epistemic communities and multiple realities are core for Long. The fieldwork situation itself is part and parcel of the conceptual apparatus; the observer–observed interaction is part of the overall 'arena' of interests and stakes in development practice. Discerning micro–macro linkages is another key capacity of actor-oriented perspectives. Long's understanding of the interdependency of various social groups and their capacity to exchange and negotiate resources enables us to seize the inside perspective of development practice. His demonstration of that conceptual apparatus seems to me to be fundamental (see section 5.3 for the theoretical references used by Long).

Long provides a clear research methodology, but does not define how it relates to development practice in general.[29] To do this, the methodology should be applied to fields other than agriculture in rural areas. Other anthropologists of development in Wageningen, at the School of Oriental and Asian Studies (SOAS) in London and at the universities of Amsterdam, Berlin and Bielefeld apply sociology of knowledge approaches to the interdependence of developers and developees that uncover the dynamics of development practice. Notably, Richards (in Hobart 1993: 61–78) stresses the difficulty of describing 'local knowledge' and uses the metaphor of performance to increase the agency of individual actors in his description of agriculture. However, applying sociology of knowledge concepts to developers wields less coherent portraits than applying them to developees.[30] Identifying interface situations (Long 1989; Long and Long 1992) between developers and developees can achieve a stronger agency focus than the performance accounts of the latter. This should also be the case in fields other than agriculture.

The strength of interpretative sociology combined with ethnographic fieldwork is versatility, and the weakness is cutting across disciplines which hamper the establishment of a school of research. While actor-oriented approaches are becoming more influential in sociology, following Anthony Giddens and Alain Tourraine,[31] they remain difficult in studying development practice. Understanding social processes within the UK or France in the age of so-called late-modernity, post-modernity or *Spätkapitalismus* succeeds, and the actor orientation proves its worth. In development practice, that worth is not at all new. In power-ridden contexts, such as development practice, social processes are particularly pertinent. Planned development interventions have modified and accelerated social processes less than colonial domination; nonetheless, development agencies voluntarily and involuntarily create and foster social agency.

A possible explanation for the slow progress of actor-oriented research is suggested by Marc Poncelet, who analysed the attempts to take culture into account in development planning. He shows that development can become a 'culturophage' (Poncelet 1994: 210–231). Klitgaard (in Serageldin 1994: 78),[32] writing for the IBRD, could be used as evidence: 'In the 1990s I believe the issue of *how* to take culture into account will take center stage'. Klitgaard wants foreign experts to act as therapists to their counterparts. Long's methodology takes culture into account, but it does not totalize[33] local culture; development practice appears in its hybrid form with cultural traits from the developer and the developee. Therefore, an actor-oriented methodology cannot satisfy the purpose for which it is put to work in a development agency, i.e. to explain the developee. The actor orientation forces research to take the developer's culture into account as well.

While the US researchers were innovative throughout the second and third development decades, anthropologists in Europe cast new light on the inherent problem. In the middle of the 1980s, European development agencies such as NORAD, SIDA and GTZ (Norwegian, Swedish and German governmental aid agencies) began to look at their use of social science input,[34] attempting to draw lessons from the USAID experience. These efforts have advanced rather slowly, possibly because of the inertia of development agencies staffed by civil servants. However, it is important to note that European and US development agencies often compete for opportunities for development interventions and copy each other's latest gadgets.

Despite the debate taking place, development anthropology has not been received in Europe. The Atlantic is wide and erudition is scarce. Furthermore, academic resistance to development anthropology in Europe is widespread and has made it more difficult for European agencies to learn from USAID. The question of whether academia's inertia is higher than that of the development agencies is not pertinent. Certainly, the academic institutions in Europe are less dynamic than the entrepreneurial universities in the USA, competing for funds from USAID. Also, foreign policy to contain communism in the Third World would have met strong reactions in European universities, where students have some influence on university policy. In fact, successful consulting work can harm an academic career, as Claude Arditi or Dominique Desjeux in France and Frank Bliss in Germany have experienced. This points to the resistance of academic schools of thought to engaging with development practice.

The conditions for anthropological research in development practice have changed through the interest in actor-oriented research. The case studies published in the 1980s consolidated the anthropological contribution on certain types of projects. They were intended to build up original contributions to a particular type of project.[35] These would be called upon for similar projects with respect to the field addressed and the social context. Nothing prevents a continuation of the 'New Directions' applications, but the actor-oriented research proposals are subject to different conditions. Can (and should) the two approaches co-exist?

The pursuit of sustainable development recycles the 'New Directions' results. Development anthropology continues to grow, if only because the donors continue to provide funds, whereas anthropology of development relies more on the ambition of researchers. While sustainable development has decisively turned into the central paradigm of development agencies, it is still not free of contradictions and blank conceptual spots. Anthropologists could be very helpful in addressing these.[36] Compared with the potential that anthropology has via the basic needs orientation of 'New Directions', its potential via sustainable development is considerably greater. Perhaps anthropologists will make better use of the opportunities of sustainable development than they did of 'New Directions'. One should look at development anthropology's limits, the fields addressed and the power differential between development agencies and beneficiaries (conditions 2 and 3) to see where actor-oriented approaches might expand, especially in order to reduce the influence of development anthropology's colonial heritage.

2.4 The debate about the object of development anthropology

Development anthropologists fell short of providing accounts of the failure of development aid. The change from development anthropology to anthropology of development is more driven by the overall failure of aid than by analytical progress. However, anthropologists should not be blamed for failing to decipher development practice. To do so, they need access to the practice. Without first-hand access, an historical account of practice, compiled from project reports, enabled Jean-Pierre Chauveau[37] and Raymond Apthorpe to assess development practice. Chauveau demonstrated that rural development projects continuously reproduce the same type of failure over several decades. But such possibilities are limited. In the end, it remains necessary to observe the interaction of developers and developees during the implementation of programmes and projects to comprehend development practice. The scarcity of research on development practice reflects the difficulty of accessing this field.

The inroads made during the 'New Directions' phase might be regarded with hindsight as a contribution to the calls for an anthropology of development, despite insufficient attention being given to the question of development practice.[38] The outspoken '*realpolitik*'[39] of some development anthropologists has enabled others to improve their critical understanding. To provide a caricature, imagine the happy social engineers as they pursue development anthropology while the avant-garde build an anthropology of development. Clearly, the two represent a rather inefficient combination. Rather like two sides of a coin, both sharing the same empirical accounts.[40] Often, the chosen objects of development anthropology were seen as 'applied', and thus inferior to pure research. However, this interpretation is increasingly appearing to be a smoke screen.

Besides the choice between development aid content and discourse (or ideology), the debate between development anthropology and anthropology of development can be situated at the level of the researcher's role in development practice. On one hand, there is the call for equal treatment of the development agency and the target population:

Allan Hoben, convinced that bureaucratic behavior was probably as rational as peasant behavior, undertook an analysis of the organizational rationale in USAID. He was able to do so only as a practitioner working in the bureaucracy, for only in this fashion would he have been exposed to that which 'goes without saying'. Anthropologists have been working for USAID for several decades, but almost consistently as outsiders, and none attempted to make systematic sense of the often contradictory, usually confusing and too frequently counterproductive series of USAID actions and explanations for them. ... I mention Allan Hoben in particular only because of the absurdity of the example: a host of anthropologists, many of whom have been ethically and politically effective, have dealt with USAID for decades yet not shared with the profession the basic research results upon which, our theory of practice tells us, efficacy depends.

(Partridge in Eddy and Partridge 1987: 230)

From Partridge's perspective, development anthropologists have failed because they have not addressed the core object, development practice.

An alternative view of this position is Arturo Escobar's critique of development discourse as cultural domination. His critique begins in a way similar to Partridge's, but he draws different conclusions, arguing that the anthropologist ought to assist the oppressed, the intended beneficiaries. Development anthropologists are in all circumstances in conflict with development agencies. Thus, he denies that there are viable objects of research in development practice. The advances of development anthropology in the USA have contributed to this critique, inspired by the post-modernist anthropology of Clifford Geertz's disciples (J. Clifford, G. Marcus, M. Fisher and others). Their call for a critical anthropology is extended into development anthropology by Escobar.[41] The impact of this criticism is not yet clear. Assessing the scientific foundation of 'culture as text',[42] one can expect the post-modernist paradigm to lose influence when the attention to political correctness is refocused. In the areas where anthropology has made most contributions, i.e. in agriculture and in health, the institutional interest can ignore Escobar, and the development agencies will continue to use the results. In Europe, hesitation towards the 'culture as text' school in anthropology limits the reception of Escobar's critique. While Escobar does not propose an alternative development practice, he falls into the same trap as many development anthropologists because he speaks for the oppressed when he judges development anthropology. His arguments are also used by those who pursue development anthropology:

Missing from most of the literature and consultants' reports on rural development are the voices of those most directly affected by development interventions – the local people ... To a certain extent anthropologists have played the role of surrogate and have taken it upon themselves to speak on behalf of the 'Other', a role that is increasingly questioned ...

(Gow 1993: 392)

David Gow, a development anthropologist like Partridge, has demonstrated that research on development practice can build upon practitioners' overcoming the power of development discourse, a possibility that Escobar excludes (Gow 1997; Grillo and Stirrat 1997).

The renewed discussion of development anthropology itself has not used actual development events as examples, with one exception, the most salient case of development anthropology during the 'New Directions': the Agroforestry Outreach Project (AOP) in Haiti.[43] The AOP was conceived by development anthropologists who possessed an intimate knowledge of Haitian agriculture. Originally having a target of 4 million trees to be planted as wood fuel, it planted 50 million trees between 1982 and 1989. The developmental success of the AOP is incontestable. Nonetheless, this example is used to dismiss the contribution of development anthropology as non-existent (by Klitgaard)[44] as well as to dismiss it as manipulative (by Escobar). That alone shows that Partridge's critique is correct; if development anthropology had substantiated its object of analysis, such contradictory interpretations would not be possible.

Perhaps another unique opportunity to study development anthropology versus anthropology of development would be to use Long's actor-oriented methodology on the Vicos Programme in Peru. Vicos has been developed by Cornell University, with funding from the Carnegie Corporation. The assessment of this development anthropology intervention is not clear according to Doughty (Doughty in Eddy 1987: 433–459). An actor-oriented research approach would certainly help to clarify how development anthropology's objects fared in the Peruvian society.

Ironically, the post-modernist critique could be positive for development anthropology if the discredit of the applied nature of research is diminished. Post-modernism correctly points to the civilization tenets in Malinowski, Radcliffe-Brown, Mead and Dumont's oeuvres. If all anthropology contains an implicit application, overtly applied work becomes more acceptable. Applied research would thus distinguish itself from classic fieldwork mainly by the anthropologist explicitly defining the application. Classic anthropologists thought that they could control the application by concealing the inherent orientation of the application of the research.[45] However, the post-modernist critique cannot approve the results of applied research because it has had to dismiss the notion of scientific truth in the first place. Indeed, anthropologists speaking for 'the rural poor' take the risk of being blinded by their power position. Post-modernist critiques of classic anthropologists' habits of Othering do apply to development anthropology particularly because of its ruralist bias. Working in urban areas helps a great deal when working on the habits of Othering of the subjects of study. As already observed on p. 14, the colonial heritage of anthropology is problematic because the social context of research is not sufficiently analysed.

The debate can also be situated at a more profound level. The relationships among anthropological research, the developers and developees can be less important than the intrinsic qualities of the results. Development anthropologists such as Horowitz, Little and Painter appear to be little disturbed by the post-modernist critique. Their results on irrigation and forced replacement (through watershed projects or dams) work in the interest of the developees, even in the most awesome political contexts. Nevertheless, they have to explain under what conditions one can be certain of the intrinsic quality of the results – or risk more post-modernist criticism. While this criticism might not have much impact on the development agencies, this can have an impact on the general reception of development anthropology, labelling it as instrumental in the power of development agencies. In turn, the general reception can reduce the possibilities for innovative fieldwork. In this case, Greeley's prediction (in Green 1986) can prove to be quite valid, not because of the reduced 'basic needs' attention after the 'New Directions' period but because of the vulnerability of the niches that development anthropologists have chosen for themselves. Furthermore, the development anthropologists' defence, insisting on their results in a particular project, is not viable because of its localism. Yes, improving a particular project is positive, but one cannot ignore the big picture, especially in a discipline that has holistic ambitions.[46]

Anthropology of development can reconstruct various types of developers and can understand development agencies, their planning modes and their project lineages. This could change the role and identity of those involved in development practice. Kathy Gardner and David Lewis (1996: 76) have pointed to this potential and conclude that 'development anthropology is at an exciting juncture'.

> Now that interpretive and hermeneutic approaches have demonstrated their capacity to persuade producers and consumers of anthropology of viable alternatives to positivism, we face the task of a 'critical anthropology' on a new level. History should have taught us that no power is more pervasive and insidious than that of the hermeneut, the authoritative interpreter of texts. And that there is no exercise of that power more dangerous than that which colonizes the texts of other cultures, especially in a world in which control over information is said to become more important than control over resources, manpower and technology.
>
> (Fabian 1989: xiii)

The near future will tell whether a critical anthropology is politically feasible within development agencies. The challenge that Fabian sees for anthropology in general is certainly valid for an anthropology of development, and there is still time to build on the insights gained during 'New Directions'.

When new research opportunities exist (as described by Klitgaard) and new methods and objects are available (such as actor orientation), then it is time to explore new ground. Urban and industrial contexts are an obvious field for an anthropology of development to verify viable objects of study. Although the case studies in Chapter 3 cannot serve as indicators because they were prepared without the development agency's consent, such case studies help to suggest objects of study that should be 'declarable' to development agencies in the future. The latent processes described are innovative because they arise from an unexplored area of technical assistance, combined with an unprecedented application of theory. In the best case, another conjuncture such as 'New Directions' might appear. My 'covert access' as a technically competent expert did not require me to declare any previous research. I did not have to write a research proposal to participate in the project implementation.[47] However, the latent processes discovered might in the future enable others to gain access to industrial projects. Understanding the past inroads by development anthropologists cannot lead to the latent processes, but, keeping that past in mind, one can explore them in the most effective manner. Besides the practice-related objectives of this book, there are a number of research objectives:

- to explore industrial projects for social processes concerning technical assistance;
- to experiment with an actor-oriented approach, looking for the observer's transformation into an actor of project implementation;
- to elaborate methodological specificities, disregarding the objects of development anthropology;
- to identify conditions required for the observations which explain the rejection of applied research;
- to assess the interpretative horizon of the participants in order to establish what can be understood by looking at an individual project;
- to assess the specific role of technology as the developmental content of industrial projects.

3 Constructing the intelligibility of the events based on participant observation

Technical assistance to industry is a new field to study for anthropology, and the analytical instruments used here are equally uncharted. All events constituting project implementation are submitted to one approach, consisting essentially of a type of fieldwork – gathering data through participant observation. This chapter examines all aspects of this fieldwork approach. Fundamental aspects are immediately introduced and citations from the events serve as illustrations. Instead of defining the method, it is more convenient to demonstrate it – thus avoiding an error-prone, purely theoretical description.

Beyond the research-related objectives, we have already envisaged an even higher ambition. The case studies should enable other developers and developees to find greater value in technical assistance in general. This requires integrating the results into the analysis. Developers and developees should not simply accept that latent processes decide the fate of the projects studied. Instead, by following the analysis itself, it will enable them to examine the latent processes in their own practice. Analysing Appui Technique and Autogeneración is an anthropological exploration of technical assistance. If such an exploration is validated and received by the particular protagonists[1] in the two cases, it can support others in understanding their own practices.[2]

The protagonists' efforts to understand their situation and their means to do so were not fundamentally different from those of an anthropologist observer. What distinguishes an observer is the privileged position of providing an interpretation. Protagonists reflect on events in just the same way as observers, and protagonists almost always seek to exploit this privileged observer position to change the course of events. This creates the possibility of understanding a technical assistance project. To explain this epistemological approach, one can compare it to a pressure cooker. A project is an ideological pressure cooker and the observer position functions like a little hole in that cooker, where some steam escapes. Being small in relation to the cooker, the hole does not alter the pressure inside, but allows the pressure to be measured. Similarly, the observer does not alter the project, but enables the ideological stakes for developers and developees to be read. These stakes include professional careers, reputations, pride, salaries, profits, market share and so on.

This epistemological approach has been pursued since the 1970s, especially with urban and industrial phenomena in contemporary France.[3] Ideally, understanding the social reality is the joint product of an observer and the protagonists of the social processes that occur in the field studied. The classic instruments of participant observation are reinforced. Fieldwork for this approach fails if it does not address the impact of participation or does not define how the analysis exceeds participation.

This epistemological approach is not specific to development aid, nor is it pertinent here because the observer was also a technical expert. Participant observation is always conditioned by the social and cultural processes inherent in the social reality being studied. This implies that a European middle-class observer (the author) was automatically linked to the colonial past and to development aid. Research was seen as another professional activity with motivations similar to those of business or development agency activity. But it is not the legacy of development aid which requires this approach. What makes this approach the only viable one in technical assistance is the intensity of the developer–developee encounter.[4] The protagonists of the cases involuntarily acted far from their social support. They took risks involving fundamental questions that would affect their lives. The dynamics of the project were aggressive and changing. Despite the aid legacy, an observation was often a singular event. The approach that was used is pertinent to singular observations. Prolonged fieldwork to gather more data later on is very difficult in technical assistance.

Following on from this epistemology, we must first look at the stakes (section 3.1), then understand the protagonists' attempts to exploit the observer position (section 3.2) and finally recognize the responses of the protagonists and the development agencies to the written results (section 3.3). While these three steps of the analysis are independent, they are interrelated and reinforce each other. Because of the protagonists' manipulation of the observer position, we can verify what was at stake for them; it is then possible to see how their reactions to the written results (this present text was sent to them prior to the publication of this book) are determined by the operational routines of development agencies and, in turn, how these operational routines can be overcome. This is the fundamental reason why we can make progress on the basic questions of 'what were the individuals doing?' and 'what were the development agencies doing?' We will finally conclude on this epistemology and the research objectives for an anthropology of development in section 6.2.1.

Where appropriate, the analogous elements from the two cases are juxtaposed to highlight the similarities. These are surprising given the gross differences in the context, as shown in Table 3.1 (Chapter 4 contains a detailed description). With a GDP of $US150, Chad is one of the four poorest countries of the world. Chad exports mainly cotton and imports all industrial goods. Without an economic base, there is no education or any health services for a large part of the population. Most employment is in the informal sector and provides no social insurance or vocational education that would allow anyone

Table 3.1 Juxtaposition of the conditions and actors

	Appui Technique (Chad)	*Autogeneración (Mexico)*
GDP per capita ($US)	150	1,830
National average life expectancy (years)	46	69
Project budget ($US)	1,000,000	600,000
Implementation (years)	5	2
Intended recipients	Artisans in the informal sector,* mainly welders and metal-workers	Engineers in industries with more than 5 MW$_{el}$ energy consumption
Institutions	French NGO/ local government	US consulting company/ local government
Knowledge	Manufacturing of agricultural machinery	Engineering design of cogeneration power plants
Foreign/ local actors participating in the TA events	Martin, Jacques, Pascal, Thomas/ Tahem, Dambai, Atula	John, Joe, Jack, Jim, Bill, Ben, David, Tom/ María, José, Ramón, Miguel, Hector, Aníbal, Rodolfo, Geraldo, Silvio, Severino, Lorenzo, Octavio, Juan

Notes
*As they were also in daily contact with the experts, the most prominent are introduced individually; these are Mohammad, Osama, Rahman and Ngerbo. Others appear only by name in the text. Appendix 1 shows a picture of them working on the prototypes which were the objects of the project. The literature concerning the informal sector is extensive.
NGO, non-governmental organization; TA, technical assistance; MW$_{el}$, megawatts of electricity in, for example, plants in the chemicals, steel and food-processing sectors.

with ambition to move beyond mere survival. The technology in Appui Technique reflected this. The skills involved are no longer used in industrialized countries. In Mexico, by contrast, education, social security and infrastructure are well established, although these remain beyond the reach of a substantial part of the population. The technological objective of Autogeneración cogeneration[5] was state-of-the-art energy engineering. Economically, historically, politically and technologically, the case studies are almost opposites. This serves two purposes. First, they represent vastly different ends of the technical assistance spectrum. Second, if no other variable explains the similarity between latent processes, the only common variable remains the developer–developee encounter in technical assistance itself!

3.1 Project dynamics generated by the actors' life-worlds

We now establish the relationship between the empirical object of study (the projects themselves) and the intellectual object proper (the developer-developee encounter). Can we simply confound them both, or can projects serve as anthropology's primary focus? The symbolic economy of the life-worlds (Chadian and French in one case, Mexican and US engineers in the other) determined the dynamics of the exchanges in each case. We will reconfirm this later by describing the idiosyncrasy of each implementation (Chapter 4). Here, we demonstrate that the different perspectives of the actors were coherent and complementary. Quotations from interviews and taped meetings confirm this.

The actors are introduced as they appear. They addressed each other by their first names, but those in the text are fictitious. The terms 'foreigner' and 'developer' are exchangeable and so are 'local expert' and 'developee'. 'Foreigner' and 'local expert' allude to their objective position as professionals. 'Developer' and 'developee' allude to their subjective positions as members of the project. 'Developer' is a rhetorical figure and efforts to decipher such a social category failed (for example Guth 1982). They are 'vaguely aligned by the virtue of their route into development' (Kaufmann 1997: 129). They are also called development experts, consultants, development cadre or advisors, but increasingly 'developer' is used in the specialized literature. 'Developee' (or developed) is the corresponding term for the recipient of aid and advice. By their willingness to participate in the observed events, these individuals were labelled developers and developees, but this is only one part of their lives. To label them further, we use the next best objective condition that they have in common: some are local and the others are foreign.

First, we must acknowledge the force of the 'imagined' that is constructed across the interface between foreign and local actors. Most of these actors felt that the figures of 'big brother', the developer's crusade, malinchismo[6] and so on were too simplistic. Nevertheless, the ideological operations that animated the characters of this psychological drama were violent and strong. Much energy was invested in these extravagant intellectual objects, in their construction and maintenance, although they were never reified by the actors. Beyond these objects lay a symbolic system that we can trace by reconstructing the points of reference.

In technological terms, one can identify an issue that limited the success of the projects. In Appui Technique, conflict of interest over the product quality control (oxcarts, grain mills) caused insurmountable disagreement.[7] At Autogeneración, the conflict concerned the quality of the data (the basis of the feasibility studies to be conducted) that the Mexican experts obtained from engineers in the plants. The Mexicans could not reproduce the technical discourse of the foreigners to explain the data, nor could the latter recognize the efforts made by the Mexican experts to work with the data. But both of these technological explanations do not recognize the underlying ideological operations that rendered these differences debilitating. Such surface conflicts could have been easily resolved, as the foreigners in both projects were aware of the novelty of the technology. However, below the surface of explicit expressions the foreigners could not grasp the cultural reinterpretation necessary by the Mexican experts to enable them to share their criteria for 'quality' and/or the significance of the various data. The local experts recognized the technical arguments, but they did not distinguish between a foreigner's attempt to dominate and the professional exigencies normal for the foreigner in France and the USA. The ideological operations born of the one's gaze upon the other produced an interface between foreigners and locals. What appeared to be misunderstandings were actually disagreements over the meaning of the knowledge exchanged. Consequently, we must first study the meaning assumed on either side of the interface and then study how the experiences of living the encounter was reinvested by the actors.

3.1.1 Foreign actors

Although there are individual differences, all foreigners shared an ideal-type of handling cultural distance and the local experts shared another ideal-type.[8] The foreigners of Autogeneración were developers displaying the will of the energizer, for modernization and for progress. In terms of this project, the strength of character of the protagonists was demonstrated in their encounter with the local experts. The professional identity of the developer is ultimately defined in the field. The incoherence that the developers perceived in the Mexicans' reasoning reinforced their professional identity and determined their understanding of the interaction. But, the foreigner is an expert before beginning his/her mission in the target country. His/her expertise is acquired and is not changed, adapted or influenced by the various contexts in which he/she works. In this sense, we can say that the foreigners in Autogeneración were first experts and then developers. Whereas the foreigners in Chad were developers first.[9] Critical distance from development aid, which experts inevitably experience, constitutes a vital aspect of the practice of technical assistance. Institutional demand in industrialized nations creates the coherence of 'expertise'. The case studies show to what extent individual experts vary despite their institutional definition.

Central to the life-world of foreign experts is the rift between the reality of technical assistance and the moral anchorage that they can construct for themselves. In both case studies, the cultural distance that was lived and suffered by the foreigners was transformed into protagonism. Their alienation from the local actors fed their determination to continue.

John[10] had been working outside the USA since 1982. After each engagement, he affirmed that the challenge of the journeys and the encounters with other cultures excited him so much that he would not return home. He claimed to have hated his local colleagues in Egypt and Pakistan (his recent assignments), but he learned to read and write in classical Arabic and Urdu. The better he knew what to expect from the local culture, the more readily he found references that permitted him to mark his cultural distance. Learning the local languages was one example. Martin[11] noted with pleasure that he had almost lived longer in Africa than in France. It had become a challenge for him to continue despite the frustration he felt towards aid and assistance agencies. He charged the agencies with the loss of integrity and accused them of forcing experts to spend their time fostering their careers rather than improving as developers by reflecting on their practical experiences.

Martin: Me, I'm not here to make blabla, I'm here to work and that's all. I say Chad, I don't know it, hmm, I'm not like those who come here for 6 weeks, go back to Europe, read four books and then make sense of it all; me, I have been coming to Chad since 1989, I say I don't know it and I don't have time to get to know it, that's it! So now, guillotine me! [Interview, 16/12/91]

Martin lamented the constraints that his professional cadre placed upon him; namely, to direct all of his energies towards the practical operation of the project itself rather than towards his local colleagues' competence. Questioning the nature of his actions was an act that rated similar to an act of the Inquisition. In this way, his role of professional developer became the sole motivation for him.

If a foreigner could succeed in understanding the local perspective, his/her protagonism would disappear because it is a function of the foreigner's alienation from the target culture. Dismantling cultural distance in order to enter the life-world of the local actor would require a restructuring of the developer's protagonism that is based on the local actor. However, such an appropriation of the local perspective is impossible as it signifies an elimination of the very cultural distance upon which the foreigner's justification as a subject, as a developer, is based.[12]

Martin was an effective expert, a professional, but he wanted to be a developer and a volunteer. He rejected the very idea of reaching the local actor. His job was to convey methods of organization, management analysis, etc. to countries and peoples in need of this technology. John learned Spanish, the language of '*los braseros*' (migrant workers) in his native California only because he considered it a language of importance in a '*world sense*'. The global and privileged point of view underlying these statements motivated both experts to seek a mission for modernization and development. The cultural distance brought to the target country and reconfigured with local experts produces this expert-privileged point of view. The operations that construct a foreigner as an expert give a coherence to the development experience that transcends the historical context of a given project and the individual's response to this environment.

If we define 'efficiency' as a foreigner's ability to explain his/her knowledge, Pascal was more efficient than Martin or Jacques. He was also a better expert because he was able to construct his own cultural distance through his devout Christianity. John was a better expert than Jim or Joe (of Peruvian and Argentinean origin respectively) because he, too, easily constructed his cultural distance. The link between subjectivity, expertise and otherness was constitutive of the developers' identities, identities manifested overtly in their life stories and implicitly expressed in their professional practice.

3.1.2 Local actors

We now turn to the local actors and start with the Chadians in Appui Technique. The local experts lived a corresponding experience: 'I do not accept this other, but I will conform myself to the other'; 'I will do as the foreigner without becoming foreign'. The symbolism of foreignness took very different forms in the two contexts. However, the local actors in both Chad and Mexico pursued their symbolic work of discovering, and distinguishing themselves from, what constituted the local for the foreigner. The logic of their work was similar. The differences among the local actors were the expressions of the relative success of their symbolic constructions.

Mohammad[13] requested the most technical knowledge because it was technical and because it was available to him in Chad. The origin of the knowledge was unimportant to him. Everything manufactured in Appui Technique, all technical reasoning, interested and motivated him. Mohammad considered my research to be an honour to his profession and exploited my presence frequently in front of the other artisans. If he had known how to write, he would have taken many notes while manufacturing prototypes with the experts.

Osama,[14] on the other hand, rejected the knowledge he labelled foreign because to him it represented a form of Western domination. The implementation of Appui Technique confirmed his fear and increased his determination not to be recolonized. On the first day of work, Osama brought a notebook but he found nothing that the experts discussed worthy of being written down. Significantly, though, he enjoyed my interviews with him. I went to his home, sat on the living room carpet and ate a West African meal. As the interview progressed, Osama acknowledged with increasing insistence that the experts' knowledge would benefit the artisans.[15] His curiously favourable reception of the interview process was no doubt due to the slant of my questions, which encouraged the comparison between the local experts and the foreigners. However, in the workshop where artisans, Chadian experts and foreigners built machines, everything changed. Having been to France, unlike his artisan colleagues, Osama could not dissociate the technical knowledge from its Frenchness.[16] His relationship to the project was troubled. He had difficulty working because he refused to participate in the transfer of knowledge and became defensive about his own approach to the work as an artisan.

Just as the subjective evidence of cultural distance helped John to operate better than Joe or Jim (who grew up in Latin American countries), the cultural distance helped Mohammad to use the encounter better than Osama, who had been to France and for whom the foreigners were less unknown. The subjective evidence was determined independently of the obvious conditions, such as Osama and Mohammad being Black Africans and the foreigners White and French. No such obvious differences existed between John and the Mexicans. Nonetheless, the symbolic constructions are always much more complicated when an actor knows more about the foreignness.

In Autogeneración, the various responses to the foreigners were similar to those in Appui Technique, especially in their individuality. Miguel[17] and Ramón[18] came into direct conflict with the gringo (a label often used for all non-Latinos). The foreigners were especially attentive to the two of them, as they were deeply engaged in the project and therefore offered the foreigners the best opportunity to make contact and progress in their work on the feasibility studies. The irony was that the foreigners placed their confidence in the very Mexicans whose work produced the fewest results. The foreigners' frustration ran deep. Several months later, the foreigners risked their professional relationship with the Mexicans by asking Miguel and Ramón to resign from Autogeneración. Having been sacked indirectly,[19] neither Miguel nor Ramón could tell the other Mexican experts how they felt, nor did the other experts want to find out what had happened to their colleagues. The symbolic work being so individual, it kept the Mexicans from talking about their experience or their intentions.

Three months later, I met Miguel for the first time since his forced resignation. During our talk, he made an indirect reference to his experience in Autogeneración in a story about a recent moment of reflection. While strolling from his new office, in a Spanish firm, to the monument to Mexico's independence, 'El Angel de la Independencía', that stands nearby[20] he suddenly asked himself: *'Why am I working here for a Spanish company?'*

That is where his story ended, as he could say no more. The hurt caused him by his dismissal, although evident in his attitude, went unspoken. I took his silence to be an indication of his will to overcome his pain. Nonetheless, his complex desire to understand, to become the foreigner, remained. He was proud to show me the business card that Bill had given him before he left. Finally, he explained that his children had teased him because he had begun to adopt the Catalan accent of his new employer. The other (foreigner and conquistador) *'se me pega'* (gets under my skin), he joked.

José's[21] experience was completely different. His 20-year career in power plant construction helped him to take charge of one cogeneration feasibility study in Autogeneración, quickly improving the detail in the calculations. Interestingly, he was also keen to learn from me; someone with no practical engineering experience who had learned from engineering manuals. And he certainly realized this even if I never said so. In our second interview, towards the end of the project, he thanked me for having been able to *'help the seeds grow*'. Proud to participate in a professional experience in which '*we learned who we are*', he felt that working with foreigners enabled him to confirm his Mexican identity. Clearly, the same foreign contribution that enabled José prevented Miguel and Ramón from affirming their identities. The symbolic distance from the gringo (the US American, despite the various nationalities of the foreigners, and the conquistador) that the local experts experienced had one of at least two effects on them. On the one hand, the distance could turn them in on themselves and reaffirm Mexican identity as their own – what they knew and what they loved. On the other hand, it could be projected negatively onto the foreigner as evidence of his/her imperialistic tendencies. The Mexican experts took opposing stances about the negotiation of the distance that separated them from their foreign colleagues, and this rendered all communication about the project difficult. In fact, they never talked about Autogeneración among themselves. Handling these foreign experts was a question so private that they could not share it, everybody had to see for him/herself.

In Autogeneración as well as in Appui Technique, the actors' ways of coping with the encounter were in the end not so different. The local actors in each case approached the encounter with an intuitive sense of the other's foreignness. Their distance from this other was the assumption to which they were called to react in order to define their own positions as students of the foreign knowledge or protectors of local integrity. The intensity of this symbolic work, confirming their identity (Mohammad and Rahman in Chad, and José and María in Mexico) or suffering the oppression of their identity (Osama in Chad, and Miguel and Ramón in Mexico), was fundamentally a constituent of the complex local attitudes towards the projects. Cultural distance from the foreigner was an opportunity or a menace, made him attractive or repulsive. The symbolic forms appearing in Chad resembled those in Mexico. But the symbolic form in one context cannot be reduced to the equivalent form in the other context. These forms are systemic phenomena, but they are not part of the same system, as we will show by looking at the dimensions of developer–developee encounters (section 6.1).

It is possible to conceive of a project of aid and assistance as an optic, i.e. a frame that permits an analysis because it follows the limits of the actors' symbolic constructions of an encounter. For this reason, a developer–developee encounter can be analysed using the events of one project. Irrespective of the fact that it is a planning unit for the development agency, a project is a viable anthropological object. Seen from outside the frame, the efforts of the actors appear to be incoherent. The subjectivity of each actor becomes visible only when a reconstruction of the inside of this frame (see sections 4.1 and 4.2) delimits the scope of its definition and provides the optic for its discovery. In beginning within and then moving outside the frame, or microsocial space, the links between the subjective realities of the actors become clearer, giving their identities more substance. The foreigners were not all the same kind of developer, nor were the local actors the same developees. The point of this study is not to construct a rigid topology. The resemblance of these encounters is an indicator of an anthropological condition of being in the world today. In response to the question posed above regarding the object of study, we can say that a project of aid and assistance is a viable object. This also allows us to study the exchanges between the actors of a project as an autonomous field of communication with a degree of independence from the context.

Furthermore, it is a very particular field because the stabilized symbolic exchanges that take place are generally unsatisfying to the actors involved. The obstacles that they encounter are the result of insufficient communication and of insufficient socially shared meaning. While the imagined other in Autogeneración and Appui Technique was always invoked by the actors to explain the foreigner or local actor present, most knew that, for example, the 'gringo' image was too simplistic. The symbolic work attempted to fill the absence of social meaning through which the encounter may be interpreted. Developers' life histories are chains of encounters with developees, where the building blocks of the life histories are leftovers of unsuccessful interpretations from both sides. The dynamics of the exchanges within a project were determined by the symbolic economy of these life histories.

So far, we can say that what was at stake for the actors were their personal and professional identities. Their interpretative horizon was far beyond the project itself, but they failed to understand it. Therefore, two elements of the project dynamics can be pursued further. First, how much of the subjective judgement of previous developer–developee encounters has changed for the actors and, second, how far were they able to communicate their professional situation at home to their colleagues abroad and vice versa? We will return to this in Chapter 5, when we study the latent processes driving the events. We have introduced the symbolic stakes here because of their importance for the methodology based on participant observation, and we can now determine how the actors manipulated my presence to change the events.

3.2 The position of an observer and the effects of observer presence

How did my presence function like a hole in the ideological pressure cooker? Visibly, I was recording the events by taking notes and taping meetings. Not knowing what I was looking for myself, the actors interpreted my interest according to their understanding of the events. Some actors discovered that my pretended (and certainly relative) neutrality was useful for them in influencing the events through rhetorical assertions. The first question to ask was, given my background and the conditions of my participation (assessed in section 2.2), to what extent was I able to aspire to remain neutral?

My participation in Appui Technique was possible because GRET was interested in my experience as a mechanic and an engineer. This interest is reflected in an article about my work in their journal (see Appendix 1 and note 37). Notably, 'Apprendre, c'est observer' shows that GRET used my results to claim comprehension of knowledge transmission in the Chadian informal sector. In Autogeneración, I was hired as a consultant to Hagler, Bailly, Inc., who saw an opportunity to increase my marketability as an expert – a marketability from which they hoped to profit in the future. These interests had much more of an impact on my participation than, for example, the fact that I was paid to participate in Autogeneración, whereas I financed my participation in Appui Technique myself. Perhaps more important projects – those that involve high economic stakes – are accessible not because of a biased use of the results in a journal nor because of the need to secure consultant personnel but because of the other strategic objectives that motivate development agencies; objectives that the researcher's results would validate. The reinterpretation and use of results cannot, therefore, be taken as a strong indication of their quality. Although it is often necessary, use of data does not, in itself, indicate scientific value or justification. Reflecting on the quality of the observer's results, one might ask 'are they applicable or not?' 'To whom and in the name of whom are they useful?' For us, the way I, the observer, was received during the participation, and why and how I participated, is more important.

If we can say that an observer becomes a kind of pawn for the actors, there are three principle ways they can use him/her: the local actors can use an observer to reinforce or to reduce the cultural distance of the foreigner and the technology, and the foreigners can use an observer to overcome their alienation from the local context. These tactics were always executed indirectly by verbal allusion to my presence, for example. A reference that would allow the actor to express publicly an opinion about my observing. In order for the tactic to be effective, it was important that the actors took the opportunity to make comments during the course of everyday interaction, as if it were a perfectly natural occurrence and not a staged announcement. Because of the spontaneous nature[22] of the comments, something was revealed about the exchanges. The actors' use of my presence introduced new parts of the project scenario. I encouraged this by answering every reference to myself as passively as possible while responding to the direct content proposed. My note-taking always remained unspecific.[23] Although I encouraged the use of my presence (the pawn's versatility), I did not create it in the first place. Such a role is determined by the project scenario; the observer fills a predefined position.

Accordingly, my presence in Autogeneración went beyond my participation as an engineering expert. When I was present, for example, the Mexican experts criticized the Mexican government. Since the project, as defined by the government, was called into question by the observation, they felt a need to distance themselves from it; hence, the recurrent theme of malinchismo that surfaced in the interviews.[24] The close attention I paid to the actors, the microphone on my recorder and my careful note-taking were services offered to the foreigners as well as to the Mexicans and, therefore, constituted a common experience for all. Ramón, the Mexican expert who had the most difficulty with the differences between the foreigners and the Mexicans, was able to use my presence as a means of understanding the encounter. Entering his office at 6 p.m., I asked him for an interview and he replied directly: *'I tell you straight away, I don't like the gringos, but I like their money!'*

Once I had suggested it, he wanted to do the interview right away. That evening, we sat in his office until 11 p.m., long after all the others had left. Having worked in Autogeneración, my observer position allowed him to express his feelings (towards working with foreigners). The next day, we had the following conversation in front of the other experts:

Ramón: Can we interview you?
Observer: Certainly, with pleasure.
Ramón: But seriously, there is something about your way of thinking that I don't yet understand, I would like to understand you.
Observer: OK.

He thereby signalled to his colleagues that an interview was an opportunity to communicate something about the relations between Mexicans and foreigners. No other remark or comment preceded or followed this signalling. The foreigner most conscious of the fact that his or her cultural differences were an obstacle to working as a team also used my ambiguous status to close the cultural gap. Two days after the previous exchange took place, John announced suddenly in front of the other experts present: *'Tom hasn't interviewed me, but has the others. I certainly hope he will interview me one day!'*

The interviews were seen as an effort to take seriously the team's difficulties and, thereby, to render them less debilitating. The majority of the experts were glad to be interviewed. Often, these were long monologues late into the night in the living room of the house I shared. Their appreciation of my presence differed according to how seriously they took my research. My method did not call into question the role of the developer in Mexico, which made my presence less threatening to the foreigners. My neutrality and the attention that I paid to what the foreigners called the *'caprices'* of their Mexican colleagues were important for the Mexicans. While for the foreigners I was part of those who suffered from the *'caprices'*, I distinguished myself from them as the foreigner who acknowledged the difficulties that the actors faced. Thus, I enjoyed a privileged position of a trustworthy interlocutor *vis-à-vis* the Mexican experts.

The position accorded to me by the actors on both projects was generally a function of my usefulness according to the stakes involved. However, we can say that their interest in me stopped short of the strategic threshold in Autogeneración, as there was nothing that they wanted me to help them to accomplish. Foreign and local actors supposed that I understood the other better, but my insights were not immediately useful. These were rather satisfying or comforting, but not useful for a specific purpose. In Appui Technique, their interest crossed the strategic threshold. To make reference to my note-taking and my microphone during meetings was to intervene directly in the unfolding of the project.[25] Even those actors who tried to manipulate me in ways that seemed to run counter to the project's objectives could be said to have benefited from my presence. Martin, for example, used me as his informant in a crisis; Mondai, as a confidant to whom he could express his need for technical help (which he hid from the other artisans); Ngerbo, as a cover so that he could hide his damaged tools; and Tahem (the one Chadian expert who, as an administrator, had no knowledge of the technology), as a mediator to give the others the impression that he could communicate with the artisans. More important, though, was the way the actors were able to make sense of their own experience thanks to my presence.

The discussions with the French and the Chadians were more closely conducted than those with the US experts and the Mexicans because in Appui Technique I was used more strategically. I was accepted by the actors as an expert during the interviews. Thus, a free and open exchange about the events was impossible. The dialogue turned on our capacity to distance ourselves from the events and, thereby, to assume the position of witness *vis-à-vis* our own experience. Chadians and French presented their viewpoints as if they understood their relationships. Such a constraint necessarily produced a particular mode of response in the interviews. The French as well as the Chadian experts also avoided abstractions (of the foreigner, the foreign developer, the big brother, the savage) and only referred to concrete facts. Neither their understanding nor the evidence were coherent and so sometimes they stumbled from one contradiction to another.

It also came out in the interviews that, for the Chadians, my presence was a reminder of their conflicts. I was still a nasarra[26] who came seeking knowledge about Chadians. The artisans did not distinguish between a volunteer, a doctoral candidate, a researcher from IRD[27] or a consultant working for the United Nations (UN), all of whom represented to them neo-colonial foreign interests. Nonetheless, my interest in their culture and their perspective on the aid assistance distinguished me to some extent and gave me a privileged status. The following exchange is a typical example of this status. Again, it was enacted as a natural exchange between experts and artisans during normal work on prototypes.

One artisan intensified the exchanges by publicly requesting in Arabic that I go and find a piece of steel rod: *'Thomas, chouf masura'*. (In order to proceed with the manufacturing process, it was necessary to cut a piece.) When I failed to locate such a piece, I returned to the group and announced in French: *'We don't have any'*. The artisan applauded, commenting in French (to make sure that the rhetoric was *shared* between foreigners and Chadians) to the other artisans: *'You see, he understands!'*

He then used another kind of steel rod to continue with the manufacturing. I had enabled him to proceed with his work. The other artisans were most impressed by this display of complicity, as they had been convinced that as a foreigner I would not respond to a request in Arabic (or in one of the many local languages). My comprehension of Arabic proclaimed my intention to enter into a reciprocal relationship with the artisans and caused them to redefine their relationship with me. My vocabulary of some fifty words was evident to everybody and I could not follow a real conversation. However, it appeared that the act of translating was much more important than the content of the translation.

Over the course of the next few days, I was frequently sought out by the Chadians. Often, they tried out questions on me before approaching other foreigners. Several of the artisans invited me to dine at their homes and meet their families (or, occasionally, their second wives).[28] I was also a cultural representative for those who wished to discuss different aspects of the French presence in Chad or the prospects (salary, etc.) for a welder in France. They invested me, a nasara, with the power that helped them to construct their image of the foreigner. They could use me to situate themselves for or against the experience in Chad by allowing me to bridge or to reinforce the distance between the artisans and the foreigners. For all except one of the artisans (Mohammad), I provided the necessary buffer to prevent them from having to form links with the foreign experts directly.

For the French, my presence was a '*driving force that helped bring people together*'. They appreciated my ability to help them define certain elements of the local reality of the project. My technical credibility lent a familiar flavour to the 'local character' and, therefore, rendered it more acceptable to them. They asked for my opinion when it was useful, although they generally regarded my research and, in particular, my complicity with the local actors as a nuisance.

The foreigners had exiled all non-technical local trade to a stigmatized sphere of indifference where they would not be forced to confront it. The mere presence of an observer, even with a very limited capacity to translate and approach the local actors, provoked this confrontation and encouraged the French to develop their perception of the Chadians. On the other hand, Pascal and Jacques considered the very idea that a foreigner should be interviewed to be odd. Personal reflections, they reasoned, were irrelevant to their neutral technical perception of the project. Pascal found my research naive because he was of the mind that voluntary help needed no interrogation, nor did it inspire serious reflection. The link that I provided between the technical aspirations of the French and the local reality nonetheless led Pascal to attribute my interest to benevolence. Although there was a general dismissal of the interview process on the part of the French, my personality sometimes worked against this rejection. Jacques understood when I alluded to the local perspective, so much so that he told Martin he had '*confessed*' in his interview. As it was, Jacques's resistance to the interview process was also moral – I could not be implicated in our discussion in the same way as him, given my ambiguous status on the project.

The Chadian experts were more passive than the artisans during exchanges with foreigners; since they followed the project dynamics carefully, they perceived or sensed the symbolic work better than anybody else. Little by little, they realized that my work actually constituted a pertinent interrogation of the project, a fact they had previously failed to grasp. Until that realization, they had simply shared the artisans' assessment of my presence and considered collaboration with the foreigners to be beyond their limits. Dambai, one of the local experts, was delighted by the way the artisans related to the observer.

Dambai: They take you as you are and this gain of their trust was really automatic. When you took notes, there were some who did not even worry about what you were writing. So, for me, this confidence ... no one was intrigued by your note-taking, by whatever you were in the process of doing. I think the exchange certainly helped bring people closer together ... very positive and less mistrust. [Interview.]

He found my presence to be useful to the artisans and perfectly in keeping with his own efforts to achieve a better understanding between them and the French experts. He was amazed that I should ask for his opinion about Appui Technique. His experience with foreign experts had convinced him that criticizing technical assistance projects was categorically impossible. His colleague Tahem noted, however, that as an expert he had always considered my interviews when reflecting on the project. In his efforts to be as much of an expert as the French, he endeavoured to be an administrator like Jacques and to be as perceptive as an ethnologist.

The attitudes of the actors towards me in Appui Technique can thus be described as both rich and complex. My research helped the actors to understand their own communication and the symbolic forces that organized it. In Mexico, on the other hand, my social role was more limited. The main reason for this limitation was probably the lack of contact between the beneficiaries of Autogeneración, i.e. the engineers in the factories of heavy industry, and the foreign experts.[29] The lack of personal contact in Autogeneración actually made it easier for the actors to talk candidly about their experience. Consequently, the analysis of this project relies more heavily on the actors' reflections in the interviews.

If we can describe the role that I was made to play in Appui Technique as both passive and active (in the sense that they exploited my presence), in Autogeneración it was purely passive. The US experts recognized that my presence offered an exceptional opportunity to reflect on the respective difficulties they had in working with their Mexican colleagues. They did not use me or make reference to my work in office meetings as did the French in Chad, but considered my reflections to be useful to everyone. During their interviews, the Chadian artisans (especially Osama, Rahman and Mohammad), for their part, made discoveries and reinterpreted their experience of the project based on them. They began to announce publicly what they had been unable to express previously. To a lesser degree, the Mexican engineers had the same experience.

In summary, we can say that an observer (the presence and the personality) had a substantial influence on the relationships between foreign and local actors in both projects. In Chad some artisans were able to rethink their reception of, and relationship to, the foreigners, as well as their own self-image as actors. The foreigners felt that the effects of this symbolic work were linked to my presence. The observer position helped all Chadian actors to see the attempts made on the part of some to bridge the cultural gap with communication. However, most were too implicated in the process to participate more actively in it. Obviously the ideological stakes for the actors allowed participant observation to produce ethnographic results. The interpretation of both projects is feasible,the events intelligible. More importantly, the participant observation can be defined for both despite the differences. The social, economic, political and tecnological contexts could hardly be more different, nonetheless, there were commonalities of my participant observation which allow us to conclude now on ethnographic fieldwork in technical assistance in general.

The failure of technical assistance is the point of entry for the observer and it defines his/her subsequent position. The failure reflects the distance between the discourse of development and the possibility of putting it into practice. The breakdown between theory and practice, which produces the conditions in which the observer as a project participant can become a pawn for the actors, is the result of an historical legacy of domination inherited by foreigners despite their desire to move beyond it. The attention paid to the observer as a virtual participant who endeavours to understand the social processes at work appears at least to offer the means of repairing this breakdown. Consequently, the observer becomes the sounding board for both sides to express their explanations for the failure.

The local actors in both of the projects expressed the belief to the observer that blame for the failure should fall on the foreigners. The foreigners, for their part, verbally distanced themselves from the image of the developer who only did 'projects' and failed to engage in real human contact. The observer position was finally not defined by the goal of technical modernization, but rather by the effort on the part of a post-colonial subject to become a successful recipient of assistance. Although it would be presumptuous to suggest that the failure to assist is *necessary* for the redefinition of the local actor, I have observed that, in the case of such a failure, the post-colonial subject can be liberated by expressing his/her views on the project of assistance. Whereas for the foreigners the failure to assist determines the observer position, for local actors, who have watched their country stagnate or, at worst, degenerate since the moment of their independence, the inevitable failure of assistance is only indirectly responsible for the appearance of the observer on the site. Their use of the observer as a foreigner and sympathizer is also clearly a function of their need to express their cultural pain to those whom they believe have inflicted it.

While technical competence was a precondition to get access to these events, it is irrelevant for the observer position. Actually, less technical competence would have reinforced my position because my answers would have been more naive. The more the observer remains neutral between the foreigner's and the local's rhetorical efforts, the more the observer becomes a pawn. The absence of conflicting interests allows an actor to use the pawn for mediation.

Having been positively present (symbolically effective), an observer allows the actors to express what they cannot express in front of other actors. The pressure in the pressure cooker originates in the colonial past and in the failure of technical assistance, but the pressure is also maintained by the fear of expressing oneself. This fear reflects the very self-esteem of an individual, the intimate professional identity. The observer's passiveness reduces this fear. When the developer–developee encounter has symbolic importance for social identity formation, for example by posing the challenge to the Chadian actors to acquire the power of developmental knowledge from the former colonizer, the observer position approaches that of a coach or mentor. An observer brings nothing to such a position but attention.

The observer position provides some transparency to the encounter and allows the wider social processes to be separated from the tacit rules of development agencies. Later on, we will see that this observer position was reinforced by the interface between foreign and local actors. The interface was produced by the actors' efforts to change their encounter. The more turbulent the interface, the stronger the observer position. The manipulation of my position by the actors will be the key to the definition of the management goals in section 6.2. Looking more closely at the interviews of Osama (Chad) and Ramón (Mexico), we will see how these actors used my position to increase the permeability of the interface for technical knowledge. This will demonstrate how a more lucid and thoughtful usage of the symbolic matter of the projects would have been possible.[30] The management goals that are identified follow what these actors attempted to achieve by manipulating my position. Because their manipulation is central to the analysis,[31] we will now review how the actors saw my position in hindsight.

3.3 Responses to the results by the actors and development agencies

We have concluded that the New Directions phase in USAID enabled us to explore development anthropology in agriculture and health, but that the acceptance of the results was arbitrary and was not logically significant. Insufficient analysis of the research conditions left isolated results. As a result, new development paradigms were not examined for respective objects of study. Keeping the position of observer in mind, we now clarify development agencies' responses to the results of this study. These responses are specific to the fieldwork approach and to technical assistance to industry.

During project implementation, the observations were palpable and became a latent part of the actors' reality. Writing these observations up after the fieldwork can produce explicit results. In this form, the results could have led to lawsuits and conflict, but I am still on good terms with all actors. To an extent, this is a confirmation of my analysis of both projects, and the actors' responses to the results extend the analysis. Based on the responses of the actors of both Appui Technique and Autogeneración, it seems that no analytical interpretation can erase the effects of the interface, or its symbolic power to construct identity. The response by the development agencies that designed the projects is another matter. We begin with the actors of Autogeneración.

3.3.1 Actors' responses

María[32] expressed relief and satisfaction upon reading a full monograph (150 pp., now section 4.2)[33] on Autogeneración. But she was unable to resolve her contradictory feelings about Autogeneración. She maintained that the foreigners dominated the Mexicans in a way that was not in the foreigners' interests. She recognized the necessity of understanding the encounter during the implementation and that with hindsight the encounter appeared even more contradictory. She was unable to move beyond the image in which the Mexicans enclosed the foreigners, and she used my results to interpret her own experience. The comfort she felt when reading the monograph came from the confirmation that the difficulties had not been her individual mistakes. She read that the foreigners had been dishonest and manipulative. But she was also certain that the reason why Autogeneración had not succeeded was that John had not stood up to Hector and Aníbal, the engineers in the Mexican Energy Agency who oversaw the project. María was convinced that John had known better, and, had he used all his expert knowledge, he could have forced Hector and Aníbal to avoid their mistakes.[34] She was as dismissive of John as of Miguel and, in the end, she concluded that '*Mexico was in bad shape*'.

John has not read my results in detail, but he approved what he did read. When I asked for his comments, he replied that his time in Autogeneración was '*the darkest days in my career*'. Jack[35] read the monograph, listened to my presentation in a research seminar and commented in writing. He shared Maria's sentiments, but with the opposite conclusions. He considered my methodology to be appropriate to the task and added that the case studies provided '*good base material*'. He repeated the same opinion of the events that he had formed in Mexico: John's benevolent efforts to play fair had been in vain and my results reconfirmed Jack's experience in Algeria, i.e. that local experts are humiliated, for example, when a foreigner uses the local language to work. For Jack, a foreign expert is pretty much in a trap *vis-à-vis* local colleagues.

Unfortunately, my results were not shown to the Chadian experts before the end of Appui Technique. I needed the agreement of Martin, who refused to distribute them. He would probably have been unable to continue working with the Chadian experts after they had read them. Given the instability of the administration in Chad, any foreign criticism would threaten the already tenuous tolerance of the foreign presence.[36] The results regarding the experts and artisans were simply too revealing for Martin to show them to the Chadian actors. Instead, he wanted to use them to prepare other experts for their assignments. In Autogeneración, on the other hand, the results were not threatening, even given the fact that John still works occasionally in Mexico today, and María and Ramón remain involved with the Energy Agency. Ramón, who avoided seeing me after María talked about my results, eventually conceded and read them.

It is predominantly the conditions outside the encounter that determine whether observation results can be returned to the actors. They are received by the actors, but the comprehension of such results can be difficult. As we have seen, during the encounter my presence was assessed and often exploited by the actors. The end of the project also reduced the usefulness of the observations. The actors read the written results as a kind of summation of the events that they lived. We can conclude from the testimonies that the results were available to the actors in a form that was peculiar to each of them and meaningful only to the extent that they corroborated the actor's memory. They were recognizable but inscrutable as results. That is to say, the actors were unable to use the analysis to reinterpret and process their experiences.[37] By the end of the project, they were so overcome by the idiosyncrasy of implementation that the events had lost significance for them. We have thus established that conditions outside the encounter can authorize or prohibit (Chad) the review of the results. The conflicts within the project did not hinder the reaction to my results; on the contrary, the conflicts facilitated the actors' reactions but rendered the objective understanding of the results difficult. The application of such results is therefore possible in principle but failed in the cases discussed here.

3.3.2   *Development agencies' responses*

For development agencies, my results called into question the very possibility of codifying and perfecting assistance methods because they demonstrated that there is no single management variable that consistently affects their progress. The failures and successes seem to be arbitrary. Working on a hit-or-miss basis (Scott-Stevens 1987: 97; Forss *et al.* 1988), development agencies are unable to isolate the elements that determine a project's fate. Since there is no mechanism in place to monitor implementation, we are forced to ask as an afterthought whether or not an attempt at technical assistance has achieved anything at all.[38] The target is either hit or missed, one cannot redefine the target nor judge only a part of a project. This is an implicit rule of the technical assistance trade. Different research methods are being studied, but the results are still inconclusive. Stories about the real rate of failure are spread only through the corridors of the agencies. Continual failure justifies the many voices that speak of the disenchantment of technical assistance and denounce its false pretences. Companies such as GRET or Hagler, Bailly, Inc. are in a competitive market and are unable to change the rules despite their comparatively solid and long-standing track records. Donor agencies determine the rules. However, for agencies such as the IBRD or USAID, day-to-day management and planning priorities leave little time for constructive feedback. SIDA or GTZ might be better placed for this purpose. Nonetheless, we can use the reception of GRET and Hagler, Bailly, Inc. as a good indicator of the ultimate reaction by those who sign off the loans for such projects.

The reactions to the results on the institutional level operated according to a simple logic determined by development industry rules. Martin suggested that my results could be a valuable tool for GRET to use to select experts. Experts could be judged according to their reactions to the scenarios and tensions described therein. But such a tool would have to be produced by a member of GRET with its institutional bias (the particular 'developer' figure),[39] one industry rule. My results would not inspire any changes in the conception of the projects because the focus is on the practice of technical assistance and not on the role of the developer. Martin's suggestion also reflects the hit-or-miss basis: using the results only to prepare experts maintains the separation between the project design phase and the implementation phase. He knew that the implementation could only be questioned during the former, which reflects another industry rule. GRET later published part of my results in its journal (see Appendix 1). This dissemination of my results was in the interests of the Chadian artisans,[40] although they were not asked. GRET also used its journal for a full monograph on a 'micro-enterprise in the informal sector of Chad'.[41] The results relating to the developee are more 'useful' for GRET than those concerning the developer. The development agencies in France and in the USA were too caught up in competing for projects to engage in the level of self-reflection that the analysis invites.

But the most important conclusion to be drawn from the reactions to the results was that it was the exploitation of my presence during implementation that made my work interesting to the actors, not the utility of my written results. Even a monograph would not have been as useful to GRET (which could not have taken full advantage of its critical possibilities) as my presence during implementation was to the actors. To put this simply, the observation results have more meaning for the actors than they have for the development agencies. The complexity of the encounter between foreigners and locals is beyond the comprehension reflected in development agency documents, and it follows logically that the observation results cannot be related to agency records about project outcome.

The inadmissibility of the results at the institutional level is corroborated by the experience of Erika Moser-Schmitt (1984), who independently published the results of her participation in an urban development project of GTZ.[42] GTZ refused to publish her research despite the fact that it was not at all incompatible with the policy of GTZ. She reflected upon the 'fear' that appeared in GTZ's resistance.

It is important to emphasize that one can reconstruct the experience on the 'inside' of a technical assistance project. The reconstruction may appear impossible for 'true–false reasons', but these can be circumvented in various ways. The caution of development agencies, or rather their resistance to qualitative social science research, makes sense insofar as the reconstruction of the evidence could indict them. The criticism is received as hostile and negative rather than constructive and useful. This is the agency's error and the basis for the antagonism towards qualitative social science. There was no political or professional reason for not returning my results,[43] either for the development agencies or for myself.

The conflicting response to the results (understandable but unusable for the actors) was not a function of the methodology. The contexts in which research on technical assistance is performed contribute to such paradoxes. The experts know that feedback from practice is not pertinent to the careers in their institutions. We must endeavour to understand better the current conditions of research so that we can decide which approaches work and are therefore likely to yield useful results. However, the fact that the results were pertinent, but invalid for the development agencies, is a reflection of the institutional rejection of the approach. By increasing the status accorded institutionally to implementation, the review and application of the results would become possible.[44] The gap between the importance that I placed on my intimate knowledge of implementation and that of the engineering objects rendered my results useless to the US consulting company. On the other hand, the smaller gap between field knowledge and the objects pursued by the French NGO enabled them to use my results from Appui Technique for their own purposes.[45]

Using the actors' immediate reactions to my results, it was possible to deduce other reasons for the development agencies' rejection. The hybrid nature of the results (i.e. they dealt with both developer and developee traits) rendered them unusable for agency attempts to learn lessons in both cases. But there was no reason for this because the hybrid nature is not an anathema when 'multiculturalism' is generally positively qualified. Where does the unusability come from in the end? Development agencies acting as the last bastion of Western arrogance seems to be an unlikely explanation.

Another possibility is the symbolic complexity of the meaning of technology. This possibility appears to be more plausible when we look at the first latent process (before doing so in section 5.1, we will sum up each case study). An account of the manipulation of technological parameters within development agencies could confirm this. Such an account is likely to show that 'the smoking gun' of the technocrat is problematic and development agencies are reluctant to try to control technocrats. Technocratic planners' disregard for the hidden character of the smoking gun may explain the contradiction between the rejection of the results and the recognition that the results are sensible.

Given the stakes for the actors, their use of my participation and the response to the results, it is plausible that the expansion of development anthropology beyond the New Directions phase was not hampered by intrinsic reasons. Despite the power differential between agencies and the 'beneficiaries', applied research can produce genuine results. Technical assistance to industry is a viable field for research and the actor orientation corresponds to the general conditions for ethnographic fieldwork.[46] This is the basis on which three latent processes at work may be elaborated (Chapter 5). Commenting on the IBRD's *Handbook on Technical Assistance* and the evaluation results of the IBRD's Operations Evaluation Division, we will then see to what extent the operational routines in development agencies reflect the latent processes (Chapter 6). At least in principle, it will be possible to challenge the contradictory responses from the development agencies with the latent processes. In so doing, we undermine the contradictory response instead of deploring it here.

3.4 Other repercussions from the fieldwork

Understanding participant observation through the actors' reactions and through the responses to the observation results characterizes contemporary anthropology. The ideological intensity of the encounter between developer and developee reinforces this. Fieldwork in any area of social reality is dependent on broad social processes. The epistemology described earlier is responsible for the viability of fieldwork in technical assistance. The specific context of technical assistance has two more implications for fieldwork. We outline these implications here very briefly, but then we will ignore them.

The role of technology in society is central to our industrialized world, where there is no longer any distinction possible between science and technology. Nuclear power, genetic engineering, cloning, climate change, artificial intelligence and many other issues constantly remind us that human values, democracy and a better future depend on the possibility of social choices in technology. What happens in technological assistance reflects this. These modern issues have made determinist analysis of technology laughable. Major technological shifts reflect profound values and hierarchies of society. Unless 'only a god can save us now'[47] from technocrats, one can take technical knowledge from one society and do something different with it in another. The question is, who can do that and how? We cannot be certain. Technical assistance practitioners do something different with technical knowledge in another society without knowing how they do that. This is one theoretical foundation of this study. We do not know whether technical assistance is in general imperialist domination or liberating utopia because technical assistance practitioners transmit values unconsciously. In admitting this, we assume that transferring technical knowledge holds a potential for social choices. Showing that practitioners actually negotiate the cultural dimension of technology, even if they partly ignore how they do it, indicates the potential of social choices.

Leaving the wider debate out of this analysis of technical assistance does not imply a normative position. A deterministic view implies the domination of importers of technology by exporters, or of developing societies by industrialized ones. If deterministic views are wrong, then this domination does not follow.[48] If on the contrary, determinist views are accurate, these research results are not pertinent. Other than wasted time, the result is nil. Therefore, overall, excluding the wider role of technology and technological change in society does not imply a normative position.

To help this analysis, we use selective works of Marcuse, Feenberg and Habermas. Although theoretical research should include differences in their positions, our application is too limited to consider these differences. The same applies to social theory. This implicit supposition follows from the above. If there is limited inherent causality of technical knowledge, what is 'in the driver's seat' in technical assistance? As we have seen with regard to the actors' life-worlds, the identity of developers and developees plays a fundamental role. At the end of this study, we will have reviewed more arguments so that we may conclude that technical assistance is fundamentally determined by the shifting identities of the actors. The participants of Autogeneración and those of Appui Technique are not the same developers and developees, respectively, at the end of these technical assistance events.

The theoretical works of Friedman, Giddens and Habermas concern modernity as a specific historical movement. The actors of technical assistance have by definition a modernizing ethos. This creates a specific identity positioning. The Theory of Communicative Action has often been criticized for having some arch-modernist implications. Identity formation being a fundamental part of technical assistance, referring to this theory could represent an acceptance of these implications. In other words, distinguishing strategic from communicative action in the encounter between developers and developees (section 5.4) could imply that their communicative competences mark levels of modernity. Similarly, Giddens's theoretical work on the self-reflexivity of individuals in society could be employed for technical assistance. The global anthropology pursued by Friedman (used in section 5.2) concerns available identity positions in a global identity space. Again, these theoretical positions are not always compatible.

Regarding both technology and social theory, selective applications do not require theoretical clarification because the quality of fieldwork is of central importance (therefore it is evoked here). The actors' faculties are more important than theoretical coherence. Ethnological analysis establishes the fundamental conditions of developer–developee encounters. Participant observation in technical assistance, as described above, can neglect both technology and social theory because the actors' faculties determine what is going on. The two implicit assumptions in this study are justifiable because participant observation is feasible. The theoretical demonstration of these fieldwork implications will be addressed later. Readers familiar with these theoretical aspects might approach the following interpretation of the events with this in mind.

4 Interpretation of the events

To establish the idiosyncrasy of Appui Technique first, and then that of
Autogeneración, we review the events, behaviour, knowledge, technology,
prototypes, exchanges and roles of the actors. As the starting point for the
interpretations has been fixed, it is now possible to avoid confusing the events
themselves with my interpretations. To highlight these, I again use the first
person. For clarity, we will later define the latent processes from the outsider
perspective, but these were purely a product of the actors' reasoning within
the insider perspective. For accuracy, we begin by establishing the insider
perspective. Originally, the actors' explanations were rather aggressive, since
then they have moderated and softened their judgement. Having observed
the actors, I can describe the degree of confrontation in their behaviour.
Sometimes they were more dismissive of their colleagues than at other times.
What we have examined up to now is already beyond the actors' understanding
of these events. This is self-evident; they had to act in the heat of their
encounter. We have already seen what was at stake and how I was able to
record it. There follows a description of the actual work; this is consolidated
by showing how the actors took part in it. The interpretation of the behaviour
is followed by comments that will help readers who are unfamiliar with
technical assistance to use their judgement.

Practitioners intuitively recognize the coherence of the events. Those
unfamiliar with technical assistance can refer back to the observer position
(section 3.2) if the interpretations are unclear. In fact, most of the aspects
assembled here require less interpretation than does the observer position.
A minor misinterpretation of the observer position will not affect the
interpretation of the events significantly. The coherence also reflects the
actor orientation and provides the empirical foundation for this study.[1]
Reducing this chapter to twenty pages has been difficult, but the reader can
reconsider this coherence when other aspects appear later on. Obviously,
consolidating such complex events into a few pages is in itself reductional,
however the insider perspective that has been established reminds
practitioners of their experience and allows other readers to assess the
progress of our results.

Before we look at Appui Technique, we can reformulate the basic question:

the fundamental logic of the communication in each of the projects, shown in the interfaces (an interactive filter), was essentially the same. However, the institutional interests, the ideological stakes involved, the technologies and other parameters within the two projects were very different, almost the antithesis of each other.[2] Furthermore, a sociological analysis of the organizations that were involved in Mexico and Chad would reveal very few common characteristics.[3] How, then, can we account for their most important and decisive similarity – the common logic of exchange? The way in which the interfaces worked answers this question. Previously, we could ask what connection, if any, this similarity had with the developer–developee encounter. Is it an effect of the meeting of an ethnocentric occidental philosophy and a native perspective? Or does such a similarity demonstrate the capacity of occidental thought to render banal and reduce to confinable proportions even the most complex and diverse phenomena that it registers as 'local'?

The complexity of the events suggests rather that the common logic of exchange is the result of a dialectic of communication particular to these events; a dynamic produced in the encounter but which nonetheless has an existence independent of the individual relationships between the actors. This dynamic feeds on the colonial past and takes shape in the actual practice of aid and assistance. In Mexico, the defensiveness towards the gringo does not, in point of fact, refer back to a colonial past, but can be read as a response to the cultural domination that characterizes the relationship with its northern neighbour. The very real colonial past in Chad was still apparent in the lives of the people that took part in Appui Technique. The French military presence in Chad had a direct impact on its inhabitants and acted as a constant reminder of French influence. Although, strictly speaking, the actors were free to create relationships based on their actual personal experiences during the projects, the weight of the foreign power that descended on Mexico City in the form of the US engineers and on N'Djaména in the form of the French experts determined to a large extent the terms of their interactions. To put it in developmental terms, both projects were essentially about a transfer of knowledge that was meant to bridge the gap between the civilized and the unsophisticated, between the technologically advanced and the less-developed nations of the world.

The local languages in Mexico and Chad constructed similar fields of meaning around the words '*gringo*' and '*nasarra*', which both accuse and implicate at the same time. The terms accuse the foreigners insofar as they represent the heritage of their nation's colonial power. However, they also imply the guilt of the Mexican and Chadian people, who are unable to receive foreigners as individuals unconnected to their culture's domination. The artisans in D'Njaména, as well as the engineers in Mexico City, engaged in important work on their identity. The element that allows us to make this connection is the figure of the Other.[4] When one actor addresses another, he can relate to this person as an individual (as an 'other') or in terms of his professional, cultural or political affiliations, about which the actor already possesses pre-existing prejudices (possibly as an 'Other'). Accordingly, we reconstruct the figure of the Other as it existed for each of the groups in the projects: 'gringo' or 'nasarra' for a foreigner in the imaginary of the local actors, and an 'obstinate Mexican' or 'roublard' whom the foreigners presumed to face. The prejudices were strong on all sides. The local actors' intimate implication when encountering a foreigner was generally stronger than that of the foreigners. It may help the reader to replace the term Other with gringo (for the Mexicans), nasarra (for the Chadians), roublard (for the French) or obstinate Mexican (for the US Americans) in a phrase which is unclear.

What the foreigners called 'the project' remained insufficiently defined for the local actors; it did not correspond to a precise activity. Autogeneración's purpose was to establish power stations, but what did that mean in the practical terms of the allocation of tasks? The goal of Appui Technique was the development of the informal sector, but one can go about creating employment opportunities in various ways.[5] If we can say that in a certain sense 'development' – or the work of 'projects' – is a cognitive and social act of domination,[6] then we can imagine the difficulty encountered by the actors in their attempts to define themselves and others in their predetermined roles. We can further hypothesize that this difficulty would give rise to the symbolic work that generated the Other in the interface; an image that irreparably altered the actors' perceptions of each other in their individuality.

4.1 Implementing Appui Technique

The experts had rented a house outside the centre of N'Djaména (the capital of Chad). Jacques and Tahem shared one office, Pascal and Thomas another, and Martin, Dambai and Atula worked in the central meeting room. Besides tables and chairs, air-conditioners were the only furniture. In the yard behind the house, the typical machines and tools of a small metal workshop were installed. That way, the experts worked with exactly the same technical means as the artisans in the workshops.[7] They bought the usual raw material (steel rods that were 5 m long) that was used, and then asked the artisans whether they wanted to manufacture oxcart prototypes and grain mills under their guidance. Most of the manual work was carried out by the artisans. Some had to do other work and only came occasionally;[8] those with no ongoing orders came every day. Suitable pieces of steel rod were cut, welded and bolted together and finally painted. The 3 months that I was present coincided with the most intense periods of activity, when five prototypes – two oxcart models and three different mills – were completed.

Appui Technique ended in December of 1995. During the 3 years since my departure in December 1991, the foreign presence was reduced. After Martin, Pascal and I left at this time, only Jacques remained in Chad, while Martin returned several times for short periods. Jacques finally left in June 1993; since then, the IBRD and the Chadian government have repeatedly tried and failed to redefine the project. The Chadian experts continued working with the artisans but they were unable to begin other activities. At the end of Appui Technique, only the five prototypes had been produced. Of the series of thousands of oxcarts that the experts anticipated would be manufactured based on the prototypes, there were none.

In early 1992, when Jacques was the only foreigner present, the artisans were in the process of making a pre-series of ten oxcarts for a commercial buyer but realized suddenly that they could not cover the cost of production. The purchase price of their raw materials was too high because the artisans were reluctant to use their cheapest (and thus scarce) source of raw material. Furthermore, the Chadian experts were generally ineffectual in co-ordinating production between the different artisans. Only Mohammad made a real effort to organize their work. Osama, Ngerbo, Aziz and Mondai had said that they would participate in the fabrication of the pre-series, but they kept their distance in the end. Others explicitly refused to get involved. We can conclude from the proposal put forward by the Chadian artisans in 1991 (Appendix 2) that the French succeeded in training the Chadian experts, who had apparently adopted the former's perspective on the project. The French would very probably have helped the artisans to start mass production if the budget for Appui Technique had allowed them to stay long enough. Finally, the Chadian experts tried some other activities, a small yoghurt factory, ice production and wood furniture, but they had little experience of these. Pascal had been working on agricultural implements for 6 years in Haiti, Jacques owned a workshop in France and Martin had taken part in similar artisan projects for 25 years. If they could not succeed in their areas of core competence, there was little scope for success elsewhere. Today, some of these artisans still occasionally pass by the yard to see whether Appui Technique might offer them any new initiatives despite the fact that all the experts have left.

Until now, we have focused on tangible results – the reproduction of technical objects and the contact between the French and the Chadians. Now we reach the intangible mechanism behind the results – how the perspectives of the actors determined the outcome of the project, an outcome that contradicted the intentions of the actors.

The French were able to train the Chadian experts to manage the production of prototypes; they communicated to them their manufacturing knowledge. The Chadian experts, in turn, adopted and, therefore, confirmed the validity of this knowledge. The artisans, on the other hand, remained out of the foreigners' reach. From the latter's perspective, the artisans were 'incorrigible' and refused to learn. To the mind of the artisan, the foreigners offered nothing that they could use. Despite their interest in improvement (at least as strong as the experts' desire to transfer their knowledge), the artisans were not capable of receiving instruction from the Chadian experts, who, for their part, could not effectively transmit what they had received. The achievement and subsequent symbolic destruction of contact with the

Other was limited to the relationship between the experts and did not extend to the artisans.[9]

This project caused a crack in the continuum of colonial presence in Chad. In this privileged space, neither the artisans nor the experts had to explain why they were there, they all had familiar roles. For example, the artisans spoke little of themselves in front of the experts. There was nothing to explain to the others, and yet, face to face, no one was equal. We may thus refer to this space as a 'neo-colonial' condition, characterized in Appui Technique by a classically reciprocal dependence between the foreigners and the autochthons who could not assume their roles without the other. The failure of the foreigners to act as effective developers and for the 'natives' to be accomplished developees may have been affected in part by the unfavourable political situation in N'Djaména, but its main cause was the actors' incapacity to establish a reciprocal rapport based on exchange rather than on dependence.

The neo-colonial condition of the project manifested itself as a mutual, unwitting silence that forbade the construction of the social meaning in the encounter, both between the groups and among the participants on each side of the interface. This silence was, to a certain extent, predictable given the circumstances of the encounter in a project of aid and assistance, but the actors also withheld information that would have been helpful or encouraging to the other: confirmation of the other's competence, constructive criticism and questions about the other's private life or health. Their reticence can be better explained as a reaction to the presence of the Other, bearing in mind the psychological influences of the political relationship between France and West Africa. The ease with which the artisans and experts discussed the French presence in Chad and their mutual support of this presence is evidence that the antagonism dividing the groups resulted from an indisputable historical reality, but it was not strong enough to produce a confrontation between them. The presence of the Other seems, then, to account for the moratorium on the foreigners' identity, the symbolic aspects of aid and assistance, as well as the local 'unknown' and other issues related to the artisans.

In fact, Appui Technique was an empty shell that each participant filled with his own meaning. For the French, it was a place where they could enact their will to energize and valorize the artisans' workshop in accordance with their development mission. For the Chadian experts, it represented one more exercise in aid and assistance, but one in which they found themselves deeply invested with the desire and determination to establish an open relationship with the artisans. For the artisans, it would have been a well-worn exercise except that this time the foreigners seemed genuinely to offer them the opportunity to benefit commercially and hone their skills. Certainly, among the individuals of each group, there were differences of opinion. In order to give more collective meaning to the project, they would have had to discuss these differences. The impossibility of this exchange gave rise to the silence which kept the shell empty.

It was only on rare occasions that an actor could exhibit his desire to understand the encounter better. After presenting the experts' proposals, Martin invited the artisans to have a closer look at them, reassuring them that *'this is not just a White idea eh!.'* He believed his propositions to be clear and convincing, but he needed to take one more step – to remove the label 'White' from his technical knowledge. For a discussion to take place, a Chadian would have had to explain to Martin that the foreigners could also remove the label 'Black' that they occasionally attached to ideas themselves. But this was impossible. Face to face with a French person, no Chadian would challenge his or her argument or make the smallest claim to the contrary. Ironically, the foreigners were discouraged because the artisans did not solicit their advice, a phenomenon that they could not understand and from which they concluded that the artisans were *'very reserved people'*. Finally, after 3 weeks of working together, a Chadian expert expressed, in a very practical example, the resentment that the artisans felt towards the French. Dambai explained that the artisans wished that the project would at least give them lunch. It was a key moment in the project's evolution. Directly confronted by the artisans' confusion about their intentions, the foreigners became defensive, condemning the artisans' demand as parasitic.[10] As the foreigners considered this attitude to be inconsistent with the role of an artisan, they judged it immoral – an act of betrayal. It should also be noted that the different perceptions that isolated the actors from each other were not readily apparent to any of them.

Although the hypothetical equality of the French and Chadian experts was taken for granted, their positions were actually very different. The foreigners had experience with technical assistance in other countries; the Chadians did not. This asymmetry reduced the conceptual repertoire of the exchanges to a few themes. In the course of the interviews, during meetings and in my general observations, three of these themes appeared most frequently: (1) the attempt to realize the theoretical equality of the experts, (2) the effects of this attempt and (3) the internalization and incorporation of the failure to realize equality in their relationship with the other. One way in which the experts actively strove to equalize their positions was by solidarity in a moment of crisis. When one member of the team made a mistake or showed signs of negligence, everyone reacted as if the problem actually lay in the workshop's dysfunctional organization. No one directly accused the guilty party, nor was he asked to alter his behaviour in order to eliminate the problem. Such a display of solidarity confirmed that all the experts shared a common interest in the project's success.

Tahem asserted himself better than the other Chadian experts in front of the foreigners. Dambai and Atula often remained silent when they participated in conversations between foreigners. Tahem was 2 years older than Atula and 6 years older than Dambai and was still a civil servant. His greatest hope, to be recognized as an exceptional interlocutor, was entirely based on a symbolic need. His position as a civil servant was so important in this neo-colonial project in which the francophone civil servants represented the French government itself that he was often confused with the State. In fact, he did not possess the means of presenting himself as an expert to the artisans, as he was simply unaware of the circumstances of their work. Dambai, on the other hand, had all the technical knowledge necessary to be an effective interlocutor with the artisans. They quickly understood this and requested in their nine-point declaration (Appendix 2) that Dambai be their advisor. It follows from their reaction to these two Chadian experts that the artisans appreciated the experts' technical knowledge, regardless of their cultural background – French or West African. Accordingly, the artisans could only establish a relationship with the Chadian experts based purely on technical know-how.

Having the technical knowledge of the French made the Chadians into experts themselves, but their Sahelian heritage and the colour of their skin did not give them a privileged relationship with the artisans. For ordinary Chadians, the Chadian experts were no longer a part of their world. Dambai felt their rejection strongly and suffered for it. The other Chadian experts had simply accepted this break with their culture. For example, when the foreigners asked for information about the local market for a product, the Chadian experts responded according to the language and economic logic of the foreigner; a logic that directly contradicted the local understanding of their economic situation. For the foreigners, this was the height of local ignorance, stigmatized in their eyes as an unjustified demand by the artisans.[11] They were very pleased with Dambai's qualifications but said nothing to him. In the same way, they did not express their esteem for individual artisans. The result of the ambiguous status of the Chadian experts for both the foreigners and artisans was that they could not act as intermediaries.

The economic interests of all of the actors coincided. The experts knew that their work would be judged exclusively by the employment created. The artisans evaluated their participation purely in terms of the number of clients it could bring them. The technical and economic analyses which determined the choice of prototypes were correct and approved by all the actors. Although some of the artisans merely voiced their appreciation of the prototypes, others demonstrated it by reproducing them in their own workshops even before finishing their work with the experts.[12] These artisans recognized the possibility of opening up a new market and their acquisition of the necessary know-how was more rapid than the foreigners ever would have imagined. They could not anticipate that Ahmed, for example, would assimilate the foreigners' technology in a matter of days. Ahmed was also unable to discuss his private accomplishment – the functional mill sitting in his own workshop – with the experts. He explained his construction of the mill to me when I visited his workshop one evening. At Appui Technique, this same act of manufacturing a mill was much more difficult to accomplish. Had Ahmed displayed his own product and invited the other artisans and the experts to comment, he would have benefited and the partnership in Appui Technique would have reached unprecedented heights.

Although the actors shared personal goals, their interpretations of the project involved other factors that often led them to react very differently to it. Martin and Osama both insisted that the French government had paid money to appease the military forces suspected of causing the recent fighting in Chad.[13] Their visions of the contemporary relationship between France and West Africa were founded on the same basic presupposition: the governmental powers on the Quai d'Orsay[14] sent foreign experts in order to boost the stagnant economy in Chad. Similarly, they both rejected the symbolism of the 'boss' or the 'big brother' in their interaction with the foreigners. For Martin, the game between the foreigners and Chadians was predictable and *'flawed ... to get us all the better'*. For Osama, on the other hand, it was simply unbearable and several times – exasperated – he left suddenly, running out through the yard. He came up against the limited margin of productive action that could be taken as facilitating exchange – a limit which was the result of the cultural distance (alterity) dividing the sides of the interface and causing misunderstanding between the foreigners and the artisans.

While this artisan sought to bridge the divide and create understanding by approaching the other as an equal, he only succeeded once. It took him 6 weeks of working with the foreigners, but he eventually made a request for credit *'from the project'*. His request gave the foreigners an opportunity to propose an alternative. Instead of the experts offering gifts, they would engage in a commercial relationship, offering their advice against a share of the product sales. Except for this one occasion, however, he never came close to discovering that he and the foreigners had very similar ideas about how their collaboration should evolve. At the end of the following working day, Osama addressed the other artisans in French in a tone that clearly invoked the power of foreign domination: *'Clean up!'* He thereby gave the impression that the artisans needed an expert to tell them that work for the day was finished, as if they could not determine that for themselves. Ngerbo tried to deny Osama his authority, pointing to Osama and saying to me *'ay, him, the robot, eh!'*. As was often the case, my presence offered the possibility of an appeal process through an indirect confrontation. Ngerbo cleverly countered Osama's authority by using against him the very language with which he had reproached the foreigners, who, he claimed, treated the artisans like robots. The progress made the previous night had been lost.[15] We will return to the pre-eminence of Osama and Martin's reasoning below, but first we define the only point of disagreement related to the technological content.

The foreigners' central concern was the technical quality of the machines that the artisans produced, for they considered standards of quality to be indispensable in opening up a market and, thus, creating jobs. The only way of guaranteeing that the work would meet the standards they set seemed to be to communicate it to the Chadian experts and place them in charge of communicating it effectively to the artisans. The nine-point statement (Appendix 2) in which the artisans expressed the will and desire to work co-operatively with the experts in order to learn from them was precisely the kind of assurance that the foreigners required. However, the statement also contained reflections based on the artisans' experiences of aid and assistance and, specifically, with nasarras. These experiences had proved to them that they could not work with the foreigner who always kept some of his knowledge to himself with the thinly veiled intention of fostering the dependence of the Chadians. It reminded them too much of their colonial past. They saw their suspicion of this refusal to impart knowledge confirmed in France's politics in Africa. When the foreigners at Appui Technique tried to dictate the artisans' relationship to the project in order to assure the technical quality of their work, the latter legitimately misunderstood the gesture as one more attempt by the French to reinforce African dependence. This central point of discord and confusion was never able to be addressed during their exchanges. If one thing prevented the perfect coincidence of the actors' interests it was the difference between the foreigners' and artisans' theories on how to achieve good results. The foreigners believed that they could impart an appreciation of strict technical precision to the artisans by using the Chadian experts as intermediaries and quality controllers. The artisans knew, however, that they could not benefit from this type of relationship with the foreigners.[16] Neither of the two groups was aware of this disagreement and it was never disclosed. This was the end result of Appui Technique, the final product of their efforts. Logically, project implementation ends with the subject matter progress being just ahead of the last progress in intercomprehension made by the actors.

This result was obvious in many other aspects of Appui Technique. We complement this result notably with other judgements between the actors and with some examples of how translation functioned.[17] Another example of the latent discord underlying the exchanges is the way the artisans applied the term 'Marabout' to Dambai. Jacques called Martin the 'big Marabout' and Dambai the 'little Marabout'. For Osama, this title given to Dambai was nothing more than a play on words that highlighted the expert's local roots. On the other hand, when Jacques used it, it became a way of stigmatizing the local perspective and, thus, the artisans themselves. While Osama's use of the term signalled a transfer of meaning (a displacement of emphasis from native character to Chadian culture), in which the Chadian expert who had mastered the foreign knowledge was given a local identity in order to appear more familiar, Jacques's usage only served to stigmatize Chadian culture. Dambai himself stated in his interview that it was not the descriptive value of the word that made it appealing, but the ideological performance that it permitted.

The language barrier in Appui Technique also reinforced the interface. Among themselves, the artisans spoke some French, African languages and Arabic, which they did not all speak with the same fluency. However, West

African Arabic does not contain equivalents for the technical French they needed to use in the workshops. Furthermore, with the foreigners, whose command of Arabic was generally poor, there was no alternative for the artisans but to speak French. Any expression in French consequently assumed an authoritative air for everyone. A comment in French, if it were coherent and appropriate to the discussion, would almost inevitably be well received and generate discussion. On the other hand, if it were malapropos, or appeared to be a *non sequitur*, it would produce an awkward silence. The interface regulated the use of the other's language in this way. For the artisans, this meant that their use of French served their purpose of '*not opening up*' (an expression Rahman used in his interview) in front of the foreigners, but it also reinforced their domination by the French.

The foreigners occasionally used Arabic words but this gesture did not bring them any closer to the artisans, who judged these expressions according to their form rather than their content. Pascal learned many expressions of politeness; so many, in fact, that the artisans ran out of responses to his attempts to use their language. Martin tried as well to use Arabic, but found himself ignored by the artisans. Interestingly, he concluded from their lack of response that his *laissez-faire* pedagogical method was successful. The exchange back and forth between French and Arabic was relatively comfortable for me, as the artisans accepted and responded to my attempts to communicate in the language in which they were more comfortable. Their response to me indicates that linguistic exchange could be used as a bridge across the interface. Since I served as the central port of all communication, the shift functioned best for me.[18]

The alleged credibility of the experts, the result of their knowledge of local life, should have given them privileged access to the artisans' perspective. They never explained to the artisans that they acquired this sensibility as experts on previous projects. In fact, they never discussed their past work at all, nor the work of GRET. They presupposed that the artisans would not be able to recognize this credibility and they saw no way of asking the artisans open questions about their performance as experts. Martin favoured self-evaluation, using the artisans' reactions as a guide, while Jacques looked for operational gauges that would indicate how to modify the organization of the project. Tahem requested quantitative indicators of their performance to establish more rigorous management practices. Over several meetings, the experts rehashed the issue of evaluation in vain because they anticipated the artisans' resistance to dialogue without admitting to the practical impossibility of discussion.

Pascal's superior pedagogical skills served to mollify the confrontations. The artisans would have been more willing to participate fully in the project if Pascal had been in charge instead of Jacques. Unfortunately, his admirably judicious pedagogy could not compensate for the absence of dialogue:[19]

Observer: They don't know who sent us here, huh? Why we are here?

Pascal: They can tell themselves that these are interested parties who, like the ..., like the missionaries, they teach you knowledge, but if these people are used to projects, well, they say, um, that's all. They have a job to do. They can think that, they can say to themselves, given the situation in this country, they are not in a booming economy and there are other countries trying to help them. [Interview.]

Only technical issues were automatically broached in conversation and could be understood by both sides. The foreigners asserted that they were in the process of transferring technical knowledge to the Chadians. The artisans called them *'professor'* or *'teacher'* despite the protests by some experts that they wanted to work with them, not just instruct them. These two complementary discourses allowed the participants to act as if they were cognizant of the other's perspective. Interestingly, though, the artisans no longer used those words in their interviews. Outside the strict boundaries of the lived experience in the yard, the projected image of the White professor disappeared. Similarly for the foreigners, once outside the workshop, they could admit that the supposed goal of the 'transfer of technology' was idealistic at best.

The artisans' recognition and acceptance of the foreigners was hindered by the absence of discussion about the experts' professional and personal backgrounds. They could not discern the connection between the foreigner and the company he worked for, or the types of projects he chose. All they saw, in fact, were the outward trappings of the foreign service agent: an air-conditioned chauffeured '4 × 4' with its own refrigerator, an air-conditioned family car, a moped or a bike. Elaborate Chadian folk theories classify professions, salaries, modes of transport, dress code, etc. of the developers appearing in Chad:

Osama: He comes for 1 week or 3 weeks and he leaves. It's better that way, but if we keep a nasarra on for 2 years, 3 years, he will eat up all the money ha, ha! Yes, we will see.

Ngerbo: But really help us. Don't stay for 2 weeks and then go. Stay with us for 3 years and continue to guide us ... 1 month, 2 weeks, you say, 'jhallas' ['dismissed' in Arabic], 'I'm off!' Then that's the part that we talk about, there. It's the expatriate, as we say, they screwed everything up. [Joint interview; for them, there is no contradiction in their statements.]

During the course of one interview, various images of the nasarra could emerge, often in the form of categorical judgements: all the nasarras come to get rich, or for some other exploitative cause. The foreigners who participated in Appui Technique were a group, 'the foreigners', and never became the subject of independent reflection. Of course, 'the artisans', 'the project', these designations had no meaning in the absence of the foreigners. Further, it should be noted, the artisans – this uneducated group of entrepreneurs – had considerably increased their social standing thanks to their artisanal work, which allowed them to maintain an economic stability that was rare in Chad. This increase in social standing was accompanied by symbolic work for the majority of the artisans who took a risk in using the experts' knowledge. If they tried and failed, they were accused of being '*informed by the White man*', but they were not credited with being his equal, i.e. modern, White, European.

The most important passage of knowledge occurred between Martin and Mohammad. In his interview, Mohammad explained his high regard for Martin's pedagogical methods. He would let an artisan make something by himself, inspect the result and then promptly offer tips for improvement so that changes could be made in a timely fashion. Mohammad repeatedly invited Martin to go on and deepen their technical discussions, but the latter refused to single out an artisan for special treatment. He preferred to give them all the same amount of attention.[20] Notwithstanding the fact that they could not go any further in their working relationship, the fact that he wanted to is a testament to Mohammad's desire to learn and to Martin's ability to teach him.

Another example substantiates this claim: the case of Rahman. He understood the experts better than any of his colleagues and spoke frequently and with pleasure about the differences between their worlds. The only one who favoured terms of engagement that corresponded exactly to what the experts wanted, he proposed that the remuneration for the experts be a function of the benefits that the artisans would reap from the new products. He also seemed to exhibit common sense that cut through the political confusion. Faced with an order for a product that he was not familiar with, he reasoned that an artisan should enlist the help of one of the experts to ensure that he does it correctly. He had, thus, conceptualized a classic marketing approach to business.[21]

The other artisans reproached Rahman for '*talking too much*'. His willingness to talk to the experts disturbed them. Osama, his professional partner and friend, had been to France. He often challenged the foreigner's point of view. Rahman spoke French less fluently than Osama, but wanted to join his colleague in his resistance. Most of the artisans reacted little to the foreigners and kept their distance. Among them, Ngerbo expressed himself best; where the others remained silent, Ngerbo sometimes acted like a prompter in a theatre, making sure everybody played his part. During his interviews, he explained to me why the local mentality prevented them from directly addressing the foreigners' discourse.[22]

The local reactions to the foreigners can be mapped along two axes (Figure 4.1), although, of course, the differences are somewhat reduced by putting them into two dimensions. The first represents the actor's position *vis-à-vis* the image of the developer. Mohammad and Rahman actively sought to

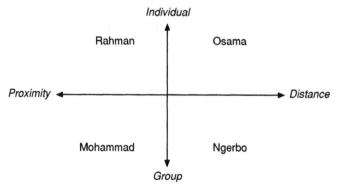

Figure 4.1 Actor behaviour

strengthen their collaboration with the foreigners, whereas Ngerbo and Osama tried to maintain a specific distance from them. The second dimension shows the actor's position on the continuum between being an individual and a member of a group. Whereas Rahman and Osama perceived differences between themselves and the experts and were ready to act as individuals, Mohammad and Ngerbo distanced themselves much less from the experts and did not want to appear to be individuals but to remain within a homogeneous group of artisans.

The complexity of the actors' responses to their lived experiences is astonishing. It demonstrates the extent to which the artisans had trouble understanding the methods and goals of the experts, who, for their part, did not present a united front. It also signals the absence of any common ground that would allow an artisan to react to a foreigner and to decipher his behaviour. Closer psychological profiles would enrich the description and might reveal how this behaviour towards foreign experts is determined. However, the dynamics of the events were such that the artisans in their diversity presented themselves as a homogeneous group in the project scenery, despite their differences. The interface homogenized these individuals. The strength of this homogenization explains the fate of Appui Technique. The encounter produced a foundational interface, which was established on three levels:

Explicit: the actors determined and accepted the specifications for the agricultural machines but did not agree with the means of manufacturing them.

Implicit: the rules of the game were put into place such that the foreigner played the teacher–professor and the artisan played the small-time contractor, but it became a neo-colonial power struggle.

Unperceived: the interface encouraged the edification of the social Other, the symbolism that imprisoned the local actors (the 'roublard' or the obstinate Mexican), and stigmatized the foreigners (the nasarra or the gringo) as a group from whom one needs to be protected.

4.2 Implementing Autogeneración

Now we can look at the second case study. The Mexican Energy Agency had contracted a US and a Mexican consulting company to assess jointly the feasibility of industrial cogeneration power plants. Whereas in Chad most of the local actors were illiterate and had no professional education, in Mexico all foreign and local actors had similar engineering degrees and consulting experience in energy. Identifying the exchange process, we will see later that this did lead to a comparable dynamic. The following description of the tasks of the eight foreign and thirteen Mexican engineers[23] only provides an impression of the context. The idiosyncrasy of the implementation provides the perspectives of the actors who created the dynamics. The observer position is used again to yield the interpretations of the encounter. Although they did not concur in their interpretations and despite their diverse behaviour and origin, the foreigners and the Mexicans constituted homogeneous groups.

The engineers rented a large office on the ninth floor of a building in central Mexico City (Polanco). One large room contained eight smaller tables, with one person working at each of them with a calculator or a laptop computer, and one large table with three PCs that were used by all. A small conference room served for longer discussions. All the work was carried out in the presence of everyone else; those who were not present were either visiting plants[24] to collect data or were involved in something else (besides Autogeneración). Two telephones were passed between the tables. A coffee machine and a water container were the only non-essential items in this office. The support staff arrived at 8 a.m., the engineers at around 9 a.m. Meetings were set for the evenings. The foreigners often stayed until 9 or 10 p.m. before returning to their hotel nearby. The Mexican engineers left around 7 p.m.; they had private lives.

Despite recently developed specialized software for the design of cogeneration systems, the design remains an artisanal process. Ben's years of experience in sugar mills and José's experience in steel mills were assets because they could judge the operating data precisely. All the engineers knew that there was no definite separation between knowledge and know-how. In other words, despite the theoretical mastery of thermodynamics, the analysis of real machines requires context-specific know-how. The difference between a successful feasibility study and a poor one is precisely this. All the engineers therefore readily pooled their know-how from the different industries they knew, since on the whole this was still a limiting factor for their work. Superior engineering performance consists of the subtlety of know-how. José, John, Bill and Tom were more productive than the others because they were able to identify the overriding parameters to optimize between, for example, a gas turbine model (which cannot be custom designed) and a waste heat recuperator (which is custom designed). The other engineers were less productive, particularly when they attempted to work out a general systemic parameter to optimize a system.

After completing its second year in December 1992, Autogeneración ended. The results were forty feasibility studies, comprising an analysis of the thermal and electrical system of a plant, and a large and a small investment in a modification. Initially, eighty studies were planned, but reviewing the results after 6 months the Energy Agency accepted that the scope of the work should be reduced to forty studies, as the quality of the results was paramount. For each investment, the payback period and the internal rate of return were calculated, a huge amount of numerical analysis. The Energy Agency planned to repeat the project with a power plant building contractor from the USA, instead of the consulting company, and other Mexican engineers.[25] However, this did not come about because it was felt that the first project was a failure. The foreigners' work in Autogeneración helped the Mexican engineers to conduct technical analyses that were previously beyond their capabilities. Nevertheless, Autogeneración was no more successful in the end than Appui Technique. The Mexicans could not recognize the technological competence of the US engineers, and the foreigners could not make their know-how meaningful for the Mexican engineers. Technically speaking, i.e. in terms of the transfer of knowledge that was meant to happen, Autogeneración did not improve local competence. Subsequently, forty similar feasibility studies were carried out, giving other engineering experts a chance.

The reasons for the Mexican engineers not responding to the foreigners and not showing the flexibility that would have allowed the foreigners to accomplish their task were deeply culturally based. The local engineers' nationalistic investment in their own logic stopped them from improving it of their own accord. They took a defensive position that caused the US engineers to be unable to communicate the very aspects of their knowledge of interest to the Mexicans. The difference in the technical capabilities of each group is not an issue here. The technology in question was so advanced that no individual engineer had enough knowledge to manipulate it easily. What did become an issue was the speed with which the engineers could assimilate and use the technology because the pedagogical method used in this project was on-the-job training. If the rapid assimilation of knowledge depended on their successful interaction with the foreigners, then the structure of their encounter must be considered in order to determine why the assimilation failed. The engineers had various explanations for this failure.[26] John, Ben,[27] José, María and Rodolfo had only produced some of what was expected of them at the project's end. But these engineers managed to maintain their productivity despite the influence of the still less efficient methods of Jim,[28] Joe,[29] Ramón and Hector. The first group's attitude prevailed over the symbolic rationality of the second group of actors; a logic caught up in the colonial past of their country. In the absence of an incontestable authority, the Mexican and US engineers should have combined their knowledge and skills to accomplish the most efficient and technologically advanced operations. In this project, the interface impeded their mutual understanding. The symbolic construction of the Other according to a colonial model prevented their concrete collaboration.

The interface functioned here like a mirror. Each side criticized in the other the very behaviour of which he, himself, was guilty. Ramón thought the others sought to dominate the project, while he was known to manipulate his colleagues. Hector suspected others of evasion, while his own technical arguments eluded concrete discussion. Miguel identified those who wanted to *'fake their way through'* in order to profit dishonestly from the project, while he cheated many Mexican engineers in the team.[30] These examples illustrate the mimetic dynamics at work in this project. As the actors in question had very different personalities but all reacted to the interface in the same fashion, we may conclude that the dynamics were larger than the work of any one person and were, rather, a function of the encounter itself. It appears that certain general concepts were common to each group and increasingly distorted their view of the other side of the interface.

Ramón was not able to find a way of working comfortably on the team. He could not acknowledge the skill of the US engineers without putting his own identity into question. When I entered his office to request an interview with him after several weeks of working together, he immediately answered *'I don't like the gringos, but I do like their money'*. He drew attention to the Other, whom he confronted in his politicized relationship with the project. Although I wanted to know what he actually felt working every day at Autogeneración, he summed up the project as an ideological confrontation with the gringo. Apart from total subjection to the 'reconquistador' or total emancipation from him, he could see no alternative. Ramón worked at one point or another with all the foreigners. His first collaboration was with me on thermodynamic models. Next, he joined John to write the electrical analysis, later visited plants briefly with Bill[31] and, finally, worked with Jack in a plant for many weeks. Unfortunately, he did not benefit professionally from these engagements. In rejecting the superiority of the *'primer mundo'*, he also rejected its knowledge. In general, any foreigner became part of a homogeneous group in the eyes of the Mexicans, no matter what their cultural and ethnic background.

The kindness that John showed the Mexican engineers was the counterpart to their efforts to demonstrate their investment in the tasks with which they were entrusted. This kindness encouraged them to foster the illusion of a collaboration. In point of fact, John found his Mexican colleagues pretentious, arrogant and vain – qualities that did not, for all of their self-involvement, make them ambitious. The foreign engineers of Latin American descent shared this assessment of their Mexican colleagues but excused these 'shortcomings' as an inevitable effect of Mexico's colonial past. In order to stand up to this Mexican pride, all the foreigners emphasized their status as worldly foreigners with international experience. Consequently, the foreign engineers born in other Latin American countries (Joe, Jim and Bill), although bilingual, spoke little Spanish with the Mexicans and refused to act as translators for them. The level of tolerance and the benevolent attitude towards the Mexicans also demonstrated some anxiety about the Mexican reaction, especially from the Energy Agency, their client. Among themselves they joked about their Mexican colleagues and vaunted their privileged perspective:

Jim:	Oh Tom, you missed the best part, Humberto came back, he was shouting at me and he yelled, 'but Rodolfo counts and he has already accepted it', so I said: 'take it or leave it!' In the end he apologised ... what do you think?
Tom:	I'd say the cultural presuppositions determine the interaction with the counterpart.
David:	And so you say, that we just react without control?
Tom:	Exactly, I would have loved to hear Humberto!
Jim:	I read a book, a guy bitching about the Bank, USAID and all the rest of it.
Tom:	Yes I know, it's called *Lords of Poverty* [Hancock 1989].
Jim (smiling):	Yes, that's it!
David:	Did you hear the latest ozone story?
Jim:	Oh yes, the story of the blind sheep in Chile, John has an impressive ozone picture in his office in DC [Washington, DC]. [Conversation during a foreigners' lunch.]

Among themselves they acted as if they had superior understanding in any field. My ethnological efforts were anticipated. The severe denunciation of all development assistance (by Hancock), even the ozone depletion in the upper atmosphere, these were all issues to be resolved by engineers. In a quieter moment, they admitted that they suffered in their work:

John:	You know as I said, I'm so glad that Bill and Ben are here now, otherwise I'd despair when you leave, whom should I rely on, you know, just to keep my sanity, it is really great that these two are good, technically yeah but also personally, um, there is some hope at least you know, sometimes, María, I don't know, I've invested too much of my time in her, but what is she doing? I don't know, I've got lots of problems with Ramón, um again, I, I don't get to him, he has these strange reactions, bizarre, um, um. [Interview at 10 p.m. in his hotel room.[32]]

John invested much energy in Autogeneración and he knew that his performance depended on his relations with the Mexican engineers. Bill, Ben, John and myself had no private or professional contact before or after

Autogeneración, but our subjective experience of our work with the Mexican engineers created strong bonds.

Joe, Jim, María, José, Geraldo and Rodolfo were all able, at some point, to acknowledge that the US engineers' methods were worth trying, although they despaired of achieving a collaboration that would facilitate the transfer of knowledge:

María: In Mexico many people, there are two sides, many people react badly to the gringos ... and on the other hand, we are very, very servile to foreigners ... that's malinchismo. It is very, very strong. [Interview.]

The contemporary form of this servility is the Mexican preference for things foreign. Only those engineers who overtly denied any communication problems at all with the US engineers did not mention this cultural tendency. The reference to malinchismo helped them to express their common as well as their individual problems in the encounter. By using the figure of Malinche to illustrate their position, the Mexican engineers established the resemblance between the conquistador and the gringo that would explain the coolness of their response to the foreigners. María refused to learn '*el padre nuestro*', just as Miguel was unwilling to be taught '*como mover el abañico*'. Catholic prayer and a woman's fan, symbols of the Spanish conquest, were mentioned during the course of my interviews with the Mexicans whenever I raised the question of foreigners. These references appeared to be part of an identity-affirming cultural knowledge.[33] In their interaction with the foreigners, other knowledges challenged their identities and threatened to reconstruct them. The US engineers knew how to determine whether or not an engineer was capable of building a plant, for example, and could, therefore, judge his worth by this criterion. For María, all interaction with foreigners was essentially the same. Miguel believed that commerce with any foreigners, especially US Americans, provoked in the Mexican a need to reaffirm his abilities. Thus, the expression of a Mexican identity independent of the exigencies of these particular foreigners informed the local engineer's response to them.

The only US engineer who allowed the Mexicans to understand him better through his openness was Bill. He spoke no Spanish and had never worked in Latin America, so he accepted and even exploited his ignorance of Mexican culture with anecdotes about the difficulty he had communicating with taxi drivers or ordering a sandwich for lunch. While other foreigners sought a close relationship with their Mexican colleagues, beseeching them to join in their effort to collaborate, Bill allowed them to be different and to resist comparison with the foreigner. What the Mexicans considered to be his typically US American way of expressing himself comforted them. I concluded from these reactions that the presence of the foreigner was less a factor in the Mexican identity crisis than their own self-perception.[34]

On the side of the foreigners, John was most determined to create reciprocally productive relations with his Mexican colleagues. Aware of the daily failure of their communication, he was very attentive to indications of the weak points in their technical analyses. When he encountered what he considered to be obstinacy in the Mexican engineers, John would retaliate with a well-known retort: *'If you don't do it, we'll bring in a gringo!'* He also made an effort to convince his colleagues in Washington of the importance of quality control in a successful project. He believed that the work of the foreigner consisted of more than a pretty written report, and that it concerned the quality of the results.

For the Mexicans, the interface reduced all individual foreigners to a single hybrid identity: the keeper of knowledge and the colonizer. Miguel produced a more sophisticated version of this image than Isabel. He and Ramón said they appreciated my knowledge but cast John as a neo-colonialist – *'con ganas de fastidiar mexicanos* [wanting to squeeze Mexicans]'. They also thought that they grasped perfectly the objectives of the project. They were wrong on all accounts and even seemed unclear about their own positions in the end. María, José and Silvio, who appreciated the technical knowledge, did not see in it a tool of foreign domination. They also declared that they never sufficiently understood the project's technical goals. In fact, only Rodolfo and Geraldo grasped the objectives and they did not participate enough to have a real impact on the whole team. It follows, then, that technical comprehension helped in the effort to collaborate, and the lack thereof worked against it. The least defensive Mexicans were those who had assimilated the important technical information most thoroughly. Ironically, José, who remained a faithful supporter of his nation's technological capabilities, was the only Mexican proud to have participated in Autogeneración. He had no trouble affirming his Mexican identity and improving his knowledge of engineering through his work with the foreigners. Despite his modesty and despite being quiet, he was the only Mexican to address everyone on the occasion of the dinner to mark the closure of Autogeneración:

José: I want to say that we all have done our best, we've all learned and we have made progress, I believe our government doesn't know what it wants, but for good or bad we've made progress for a better Mexico, and indeed it will be useful, we've all contributed. I sincerely, I don't know whether I'll continue with that work, but, in any case, I am proud to have participated. But one thing I want to tell you, John, your impression of Mexicans is mistaken, we are not as weak as we pretend to be! [In a popular restaurant with traditional Mexican food.]

John smiled and nobody said another word. This was the symbolic end of their encounter. John did not understand what José meant. José was saying that the foreign engineers' behaviour had impeded the Mexicans with contradictions; pretended weakness was only one and was rather an unsuitable example at that. But José had no better words to express himself. During the interviews with me, he used other examples. From José's perspective there had been no solution to the contradiction of a 'Mexican developee' faced with a foreign developer. From the other side of the interface, from John's perspective, there was simply no contradiction. John knew that he had tried every way imaginable to encourage his Mexican colleagues to share his engineering work without ever getting a clear response.

In order for the team to have succeeded, all the Mexican engineers would have had to be capable of José's understanding and critical distance from the political charge of the knowledge, but some of them were not. The limits of their understanding often prevented some Mexicans from making use of the foreigners' knowledge as well as their own. The limits of their comprehension also prohibited the foreigners from cutting through the bureaucratic 'bullshit' to reach them. Aníbal, in particular, reproached John and others with not having truly applied themselves to the task of aiding and assisting the Mexican engineers. He would have liked the foreigners to request more competent Mexican interlocutors, for example, when those who had started with the project failed to learn:

Aníbal: Believe me, personally, if it's necessary I tell you, really: I agree! If the people behind you cannot read a steam table, we're lost, but see, you must understand me, how I see the problem, believe me, I accept what you say, I'm not stupid ... between us, for a national energy policy, we need a lot of expert advice, people who tell us what's the best, and really, John, without being critical, he should have imposed himself so to say, not to come and say, alright you want this and I give it to you. [Interview after the end of Auto-generación.]

He had spent months simply repeating every calculation made by the engineers in order to ensure that nothing of their knowledge would escape the Energy Agency. Even so, at the end, he had not learned to produce such a feasibility study himself. Having tried the hardest among the Mexicans, he still suspected that the foreigners had hidden something.

Even the Mexican who worked longest and most intensively with the foreigners did not overcome her prejudicial attitude towards them. María, who wanted very much to be accepted into the group of 'experts' (read: foreign experts), perceived in it a tendency towards domination. This disenchantment lasted for the duration of the project.

Although the spread of engineering knowledge was the official reason for their communication, the Mexicans read their interaction with the foreigners in terms of their country's political and economic circumstances. Interestingly, in their private interaction, the respective heads of Hagler, Bailly, Inc. and the Mexican Energy Agency were able to overcome the obstacles that blocked the progress of their engineers. Neither the supposed weakness of the Mexican nor the threat of the reconquistador obfuscated their exchanges.[35] Their effective communication was achieved, and it was not the result of their forced complicity or mutual understanding but of their remoteness from the events and, thus, from the interface between foreigners and Mexicans.

In the previous summary of Autogeneración, we focused on the effects of the interface and the powerful cultural imaginary maintained by the actors. If this evaluation has been less structured than that of Appui Technique, it is because of the nature of the interface in Autogeneración. Here, the actors could act on it, ultimately limiting its coherent articulation.[36] The core disagreement between experts and artisans of Appui Technique about how to maintain quality control was not apparent. For Appui Technique, the analysis arrives one step ahead of the actors. The key transformation in Autogeneración was part of the interface: when a foreigner took a pedagogical approach (according to his sensibility of pedagogy), a Mexican engineer considered it to be a kind of deception. Presenting technical acumen in a simplified manner became in his/her eyes a covert way of keeping him/her in a state of dependence. This information was part of the interface as it was actively negotiated.[37] There is nothing of Autogeneración beyond this unresolved negotiation. Also, the technical substance of Autogeneración was more difficult to isolate and label as being local or foreign than the scale drawings and manual skills involved in the prototypes in Appui Technique. Nevertheless, despite the subtle discrepancies, the analysis of the obfuscatory effect of individual and social imaginaries on the events yielded very similar results in Chad and in Mexico.

Very different aspects of the events, behaviour, knowledge, prototypes, verbal exchanges, translations, changes in the roles of the actors and more had to be assembled to highlight the coherence of the events. None of these aspects alone can explain this coherence, which is wholly symbolic. We can now go back to the comparative approach used before – juxtaposing common elements to distinguish the independence of the events from contextual factors.

The absence of lessons learned, already mentioned in the introduction (section 1.2), should now be even more striking. Neither the borrowers nor the lenders (local governments and the IBRD) nor even the employers of the local and foreign actors were able to draw conclusions from what had happened. We can now identify three latent processes in order to facilitate the interrogation of such events by these institutions. These processes do not concern the planning of these events, nor the political context. Certainly, the IBRD and the expert employers acted within their own institutional logic in Chad. The Chadian administration never had an opinion on policy regarding the informal sector. In contrast, the IBRD and the engineers' employers in Mexico faced a strong and well-enshrined Mexican administration whose institutional logic prevailed over foreign influence.

Appui Technique was an alien exercise whereas Autogeneración was all locally determined, but both were well adapted to the local economic realities and these exercises were and are reproduced. This reflects the hit-or-miss mode of operation in technical assistance. When it was not possible to learn any lessons in either of these cases, it was possible that the latent processes originated only in the encounters between these actors. Because implementation is so removed from the organisations involved, the latent processes do not reflect differences in planning or direct political interests. If these institutions never take responsibility for these events, it is likely that they are, primarily, blind.

5 Latent processes in technical assistance

The project dynamics examination (section 3.1) showed how the legacy of technical assistance (TA) influences actor behaviour. Foreigners and local actors constructed different reasons for their work. In this chapter, we interpret this from the outsider perspective, charting these reasons. Once the latent processes are defined, we can trace them in evaluation studies (section 5.5) and then in management (section 6.3). By this means, we show how the latent processes are involuntarily strengthened by development agencies. Paralleling the operation of these agencies, we apply the latent processes to the actors observed. Using the observer position (section 3.2), we identify their rhetorical facilities to affect the encounter (section 5.4) and then ask what key actors sought to attain (section 6.2).

First, we have to separate analytically the technical knowledge from the events. To do so, we verify whether the experts failed to understand that the same instrumental core of technical knowledge can support different sociocultural ends. Theoretical research on technology by Marcuse and Feenberg confirms that the experts in the case studies were typical in that regard. This explains how technical knowledge can acquire a surprisingly uninstrumental life of its own by virtue of passing from one society to another. That passage adds another layer onto the project dynamics already observed in the actors' life-worlds. Friedman's research (Ekholm and Friedman 1995) shows that the project dynamics deriving from the exchanges of knowledge are an expression of wider social processes. This will then help us to describe how the intersection between the foreign and the local perspective constituted an interface. These three processes – the technical content, the dynamics of the exchange of knowledge and a specific interface – are sufficient to explain the events.

Until now, we have often commented on the interpretation in our analysis, thus rendering the interpretation visible. Having outlined the symbolic coherence of the events, these comments can now be reduced. Through applying theory we categorize the events. Citing expressions from the actors now serves more as an illustration than as an argument. Having established both the method and the empirical basis, we elaborate the principal results. Accordingly, the complexity of actor behaviour appears to be considerably reduced. We do not induce the nature of the latent processes through juxtaposing the case studies nor do we deduce from theory. Instead, we use a combination of case studies and theory to define the latent processes. More rigorous tests of the latent processes are a future stage. For now, we use TA documents from many other events (section 5.5) to see what we have gained.

The theoretical considerations have never been applied to industrial TA. With the exception of global anthropology, they have been developed for industrialized societies. The empirical evidence is selective and more would be preferable, but the latent processes can be identified using different approaches. Future research on the latent processes, using different bodies of literature and different bases for comparisons, will consolidate and improve their definitions.

We have seen how the practice is caught up in the problematic of the act of assistance. Projects continue to be undertaken or re-established because they respond to local and international politico-economic demands, but the envisaged objectives remain beyond the capacity of their actors. In contact with local actors, the foreigners' identity is partially determined by the teleological forces underlying their purposes. Local experts are caught in contradictions. The events analysed were unique phenomena[1] and the meaningful outcome of the encounters between these actors was their efforts to understand each other. At this point, we could either content ourselves with this result and hypothesize on its implications[2] or we could ask what aspects of these events, despite their unique nature, are likely to occur in other encounters. The first option, consolidating the basic finding, has the merit of focusing attention on the basic shortcoming of TA. However, to shatter developer complacency, more evidence on TA will be necessary.

The second option has the merit of consolidating not only TA's shortcomings but also the actors' realizations. Obviously, this choice is more constructive. If the planning habits in development agencies change, then the actors' realizations would surface. 'Latent' processes can be modified by the actors, even if such modifications do not yet enable them to change the habits of development agencies. We call latent what can occur in all TA events. Obviously, two case studies is a small basis for wider claims, but without similar case studies in the literature we must extend the analysis and then seek other ways to assess it.

5.1 Mutual appreciation of technology and the content process

The observation of TA encounters is analytically limited to what was happening during implementation. For one TA project, this focus could be a particular observation on economics, e.g. the role of cash income. For another project, the focus might be the natural environment, e.g. the behaviour of species. What happened in the two cases studies had little connection with technology[3] itself. Industrial contexts imply rather complex machines, such as welding machines or thermodynamic models. Experts in their application share learning processes implicit in them. Excluding the technical objects from the analysis is possibly appropriate in much of TA undertaken in industry. In fact, in cases where the actors always use technical arguments to achieve something non-technical, excluding the technical objects is a precondition for understanding the events. Experts and expert knowledge (especially context-dependent know-how) are treated as they are constituted or modified through the events, but the technical objects (especially embodied knowledge) are accepted for what they are. Their physical and functional conditions are not an essential part of the analysis but remain external. Observing TA thereby draws the epistemological boundary around the actors' horizon. Indeed, no expert in Autogeneración or in Appui Technique thought of creating a new piece of technology. Leaving physical and functional conditions aside is the last methodological choice we make.

Locals and foreigners in each project shared a common perspective on the basic technical import of the knowledge they exchanged. In general, there was mutual appreciation of the knowledge and its significance was agreed. In the case of Appui Technique, the artisans were already familiar with the machines that the foreigners introduced in the manufacturing workshops. They had produced these machines (oxcarts, hammer mills, hulling devices) before, but the limits of their knowledge prevented them from being able to perfect them and achieve optimum performance from them.[4] In fact, using the scale drawings that the foreigners offered them represented an advancement in the artisans' production methods. Both the artisans and the experts recognized the improvement of manufacturing techniques as the aim of the project and could therefore appreciate the new knowledge. The question is whether or not they appreciated this technical advancement in the same way.

Two examples illustrate the shared sense of how the technical knowledge should be used. An artisan made reference to the complaints of the farmers about an oxcart model. The model's weakness was in the proportion of the loading surface to the iron armature that formed the chassis under this surface. The foreigners used their experiences in different countries in West Africa, where oxcarts were already made locally, to help the Chadian artisans correct this defect in their model. The artisans accepted the suggestions because they felt that the Senegalese, for example, had superior methods of manufacture and that they should, therefore, use them. When the foreigners gave them the choice of scale drawings, life-size models or detailed sections of the oxcart drawings, the artisans chose the life-size model. They reasoned that the model would produce the best results. The foreigners unanimously agreed that this was the best choice, as a working model would clearly demonstrate the required ratio of the surface to the chassis and would permit the manufacture of oxcarts with interchangeable parts.

The second example concerns the construction of wheelbarrows. Jacques had given an artisan, Mohammad, a scale drawing of a wheelbarrow and his workers attempted to produce it. The result did not meet Jacques's expectations, who declared it '*too ugly*' to be sold. Several weeks later, another artisan explained that he (the artisan) would profit from his participation on the project by endeavouring not to produce 'ugly' products. He was familiar, he added, with the strict exigencies of the buyers, since he had already constructed improved versions of other devices. The artisan had exactly the same verbal reaction as the foreigner to the incidence of poor workmanship, although they had never discussed production criteria together. Everybody knew who were the best welders, namely Mohammad, Osama and Rahman, and both the foreign and the local experts lauded their dexterity. The experts would never have dared compare their own welding skills (I provided some good laughs trying myself). Jacques knew that Mohammad would find out how to weld a marketable wheelbarrow. Jacques and Martin recognized the value of workmanship, which increased their resentment of Osama for constantly breaking drills[5] while manufacturing a prototype of a hammer mill. Jacques and Martin were certain (correctly so) that Osama knew better. Other artisans tried to convince Osama to stop breaking the drills, but he continued his mute protest against the quality of the exchanges between foreigners and Chadians.

In Autogeneración, the knowledge to be transferred also represented a technical advancement that could be realized in practical results. Rather than prototypes, these were system designs for the production of electricity (turbines or piston engines combined with heat exchangers to use the waste heat). In order to put the technical advancement into practice, the team of experts had to calculate and predict the thermodynamics of the proposed system designs. The Mexicans had previously made similar calculations, but expected from the foreigners, who had profited from the experiments over the last few years in the USA and Europe, a superior conceptual framework that would produce novel results. The different members of the team had a unanimously favourable response to the knowledge that the foreigners brought to Mexico. The quality of the results (a sensitivity analysis of the economics of the proposed generation systems) was finally judged according to the precision of the calculations. However, the importance of the foreign knowledge lay not in calculations themselves (which were self-evident) but in the technical superiority, evidenced in the feasibility criteria.

As the Mexicans had different levels and types of experience, they did not always share the same perspective on the technical knowledge. The two who had the most practical engineering experience immediately recognized the analytical superiority of the foreigners' calculations. They admired John's competence and sought to assimilate the knowledge he offered. Discussions about their somewhat hasty calculations always ended the same way: with John reminding them that it was especially important to consider only those details which would affect the end result. Thirty years into their careers, it was difficult for some Mexicans to change their habits. However, they remained open to new ideas and their experimental nature was greatly appreciated by the foreigners. Their attitude was particularly helpful when the foreigners questioned the pertinence of certain procedures in the factories. The Mexicans questioned neither the legitimacy nor the accuracy of the foreigners' criticisms.

Some of the Mexicans, having little practical experience, evaluated the imported knowledge in terms of their theoretical background. They were greatly skilled with computers – as skilled as the foreigners – whereas the experimental engineers still used paper and pencil to make their calculations. The latter were expected to lead the team in the more repetitive tasks. However, the practical application of the calculations was not always clear to them. With the exception of Ramón, they tried to follow the foreigners' directions with a view to understanding the ultimate pertinence of the calculations. The new method was close enough to their own to inspire their interest in expanding their understanding.

In both projects, then, the technical knowledge was shared on the basis of past experience, but the actors' reinvestment of this knowledge took different forms in the collaborative work. The main reason for these differences was neither the actors' command of the knowledge nor the nature of the knowledge itself. One possible answer to why their reinvestment took different forms lies beyond the scope of the present study, namely that what appeared to be differences in technical mastery were actually the effect of differing access to professional information. Even if the Mexicans had begun with the same technical knowledge, they would not have the same opportunities to exercise and expand it as the engineering experts from the USA.[6] The foreigners' companies simply generated much more information than those of the Mexican and Chadian experts and the foreigners accumulated more know-how.[7] Although the ways in which US and French companies performed at a higher level were never quite clear to the Mexicans, they never doubted that it was the case.

The claim that local actors can share the foreigners' understanding of their technology, as I observed that they do, contradicts the classic conception of technology and culture as mutually exclusive categories, wherein technology is a threat to culture and culture defends itself against technology. This classic opposition has already been refuted by science and technology sociologists. They showed that there is a 'seamless web' between technology and culture, whether it is in light bulbs or ballistic missiles.[8] Possibly, TA is the last resort where the classic opposition is still seen as pertinent. What the actors had in common was a sense of the utilitarian essence of the knowledge – an essence that necessarily precedes and excludes the cultural meaning that technology acquires in its application. This shared instrumental significance did not, therefore, enable the foreigners as developers to transmit the ideological significance of technical modernization. From the perspectives of both the local and the foreign actors, the technical objects existed as potential tools destined for practical employment. However, the significance of their material ends differed according to the context of the actor.

In its movement from the world of the foreigner to that of the local actor, the knowledge acquires additional meaning. The passage is problematic because it occurs as a function of the identity that each side ascribes to the other. Neither the way in which sociocultural meaning limits practical usage nor the purely instrumental aspect of the knowledge is necessarily apparent to every actor. In fact, the sociocultural meanings *appear* to be generated directly from the utilitarian essence of the knowledge. Thus, it becomes impossible for the local actor to determine which elements in the exchange are the result of sociocultural biases and which are the effects of the object's instrumental significance; hence, the *ideological force of the technology*. Technology will appear to be shared by both groups once the sociocultural differences have been recognized, but its meaning will have shifted in the transfer.

Furthermore, the intrinsically instrumental aspect of technology does not logically determine its practical application. For this reason, it remains possible that the utilitarian essence of technology can serve different ends, even if these goals are quite different, having only a limited number of elements in common. The importance of knowledge finally resides in the ideological use that a society makes of its technology. Section 5.2 suggests that, according to the evidence gathered, all technical knowledge is essentially transferable because it is instrumental. However, in order for technology to move across cultural lines so that a local expert can make use of a foreigner's knowledge in his own environment, a reassessment of the sociocultural ends of that technology necessarily occurs.

In Appui Technique and Autogeneración, the foreigners shared with the locals an understanding of the practical terms of the technical knowledge – scale drawings for oxcarts, thermodynamic calculations, etc. – but they did not have the same sense of its application. In most cases, the disagreement as to the significance of a material goal was due to the origin of the knowledge, whether technical or otherwise. Often, the locals appeared to the foreigners to make use of the technical knowledge in an inexplicable or inefficient 'local' manner. This was because the foreigners could not follow the shift in sociocultural ends that had taken place when the technology was assimilated by the new culture. These differences were generally resolvable on the surface, but the discussions were not sufficiently productive and did not lead to greater understanding. Figure 5.1 is appropriate for the exchanges of knowledge in both case studies. When the actors talked to each other, they could not find out whose knowledge was in which ellipsis.

Figure 5.1 reflects the exchanges in my notes and tapes and provides a coherent frame of interpretation. The actors intended addressing the core but ended up responding to the other's image to varying degrees. Repeating this between different actors leads to exchanges where the actors anticipate and accept the necessity of making unjustifiable claims, declaring sociocultural ends to be part of the instrumental core. This is an anticipation that an actor would avoid in his/her original professional environment. This repetition becomes a process central to the implementation of a TA event.

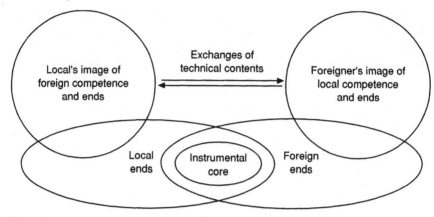

Figure 5.1 Latent process due to technical content

Declaring cultural images as being context independent and instrumental to know-how became a vicious circle. As an abstraction, the empirical accuracy of the process is less significant than its theoretical coherence. During implementation, the process gives the exchanges a logical direction. We will now confront the process with theory.

The specialized literature shows the extent to which it is difficult to separate the arguments based on an imperialist, neo-colonial model of underdevelopment from those based on a model of assistance whose goals are development and liberation. It is necessary for applied research to acknowledge that it is possible for the same technical information to promote or impede development depending on the context (the actors, the economic circumstances, etc.) of a particular project. This latent process allows precisely this. Johan Galtung (1979: 6) assumed that 'underlying technology there is a certain cognitive structure, a mental framework, a social cosmology, serving as the fertile soil in which the seeds of a certain type of knowledge may be planted'. It is still an open question whether this will lead to a withering away of TA, as he suggested. It probably will not if its agents are not agents of one homogeneous cosmology.

Typically, during the first stage of import substitution[9] such as in Chad, the artisans, having no theoretical education, induce technical conditions from using technical objects. This allows them to improve their manufacturing skills. Without theory they cannot assume that during welding the slag rises above the liquid metal, for example. Instead, they develop rich folk theories which achieve the same result. But there are more profound technical modifications that they cannot perform, such as structural adjustments to an oxcart chassis. Nonetheless, the improved oxcart model did not change the relationship between the Chadian farmer and the artisan. The only artisan who had attended an Islamic high school,[10] the Madrasa, never posed a question to a foreigner, as he assumed that know-how passes implicitly. Then there were the local blacksmiths in N'Djaména, a closely knit caste with whom the artisans exchange raw material. The blacksmiths' knowledge has distinct cosmological status (sorcerer, kingmaker, outcast, etc.) in West Africa.[11] The head of the cast in N'Djaména determines which manual gestures are permitted with blacksmithing tools. The artisans used different registers of knowledge, including inductive, Islamic artisanal and blacksmiths', all far beyond the foreigners' imagination. Mohammad, who asked the most questions and learned the most, was also a sorcerer.[12] His choices between different registers determined what kind of knowledge he applied. Those artisans who were less creative in switching between registers participated less in Appui Technique.

The situation in Autogeneración was quite different, but the same schema of technical exchanges still applied. The nature and status of data from the plant operation were unequivocal. Steam temperature and pressure are measured and translate into enthalpy and power. Nobody has direct experience of what a megawatt is nor uses bodily senses to look at a steam or a gas turbine. Among foreigners and Mexicans, the engineering experience varied significantly. Ben had been working in many sugar mills all over Latin America, I had never seen one; José had overseen the construction of small power plants for 20 years, Ramón and Lorenzo only remembered from reading books. But all would draw the same conclusion about a cogeneration system with a heat rate of 6,000 kJ/kWh: it was a very efficient system.

In both situations, the instrumental core to technological knowledge was available for the actors to refer to, if they were able and chose to do so. Herbert Marcuse's classic study of modern technology presupposes that a liberating technological evolution is possible under the right political conditions (Marcuse 1964). The instrumental core is available even for contradictory ideologies. Furthermore, Marcuse and most other theoreticians of technology demonstrated that the denial of anything transcendent is the ideology of modern technology. But this denial cannot enable the developers to reduce the role of culture, even for themselves. They are certainly carriers of this technological ideology, even though they are constantly experiencing resistance to the denial of transcendence. Marcuse referred to his philosophical teachers to demonstrate that it is the application of mathematics to life-worlds which expands this denial. All of the developers I observed did precisely that in their everyday efforts. While Marcuse's model does not fit advanced industrial societies any longer, foreigners in TA seem to be the last agents of the one-dimensional man,[13] whether they work in Chad or in Mexico, with illiterate artisans or with industrial engineers. As Marcuse showed, it is individual identity that orients technology's ideology and the same technology can serve quite different, and not necessarily cultural, belief systems. The identity of the developers is both a problem and an opportunity. Marcuse's successor, Andrew Feenberg, shows precisely that in his description of 'subversive rationalisation' (Feenberg 1995). Feenberg describes the possibility of using the transfer of Western technology in democratic ways, particularly because an agent of the import can foresee the different consequences of a particular technology. This suggests finally that no single characteristic of technology sustains neo-colonial domination or empowers endogenous processes. Nonetheless, although the actors shared the core, their exchanges ignored it, and, if the foreigners imposed their perspective on the artisans, they did so because they had only technical competence. This despite the fact that they all agreed the technologies were adequate.

This latent process is caused by the difficulty in distinguishing the instrumental core from the sociocultural ends of technical knowledge. The process explains the apparent contradiction between local and foreign experts confronted by each others' know-how while at the same time agreeing on each others' accuracy. The more ardently such a process is engaged, the better an encounter reflects what theory predicts. Only when the experts grasp the relativity of sociocultural ends does this process stop. But their professional training and experience prevents this. This process is thus the result of the nature of technical knowledge which explains the correspondingly frequent occurrence of the processes.

5.2 Project dynamics generated by the exchange process

With this first latent process, we have confirmed that all actors agreed upon and potentially shared the knowledge they were to utilize. In Mexico, all of the engineering experts appreciated the concept of cogeneration, which constituted the basis of their work. They understood that cogeneration enabled them to attain certain objectives, such as increasing petroleum exportation, energy conservation in the existing power plants, the reduction of pollution, etc. In Chad, the importance of the prototypes was unanimously recognized. The artisans understood that the local manufacture of machinery would reduce importation, create jobs and increase productivity in the workshops. The communication among the actors revolved around turbines or grain mills, but making them meaningful was difficult as their functions were already symbolically overdetermined.

This overdetermination was, of course, different on each side of the interface, thus rendering the fate of these technical objects uncertain. The passage of the technical object from one cultural context to another was thus fraught with confusion; a confusion that partially determined the context of the exchanges. Developer–developee exchanges take place in an increasingly interdependent world, in a 'global village'. TA occurs in a global sphere that it shares with world trade, tourism, global financial markets and globalizing consumption patterns. Developers sometimes complain that TA is 'crowded out' by these other stronger processes, in other words the fruits of TA are constantly being overpowered. More likely though, TA is in need of adaptation, not only to economic conditions but also to cultural conditions. Jonathan Friedman is one of the most influential anthropologists working on the global scale; the following introduction shows how he approaches his problematic:

Globalisation and Localisation

Salman Rushdie has gone underground! From 1970 to 1980 the population of North American Indians increased from 700,000 to 1.4 million including the creation of several new tribes. The world network of stock markets is overcapitalized and lodged on the fluctuating brink of the threatening crash of 1990. The governments are there to stem disaster, by means of massive credit, whatever problem that may solve. In the East bloc, large-scale ethnic mobilization threatens the monolithic face of empire while presenting new and even less manageable problems. The same T-shirt designs from Acapulco, Mallorca or Hawaii; the same watch and computer clones with different names, even Gucci clones; the nostalgic turn in the tourist trade, catering to a search for roots, even if largely simulacra, and the Western search for the experience of otherness. Ethnic and cultural fragmentation and modernist homogenization are not two arguments, two opposing views of what is happening in the world today, but two constitutive elements of global reality. The dualist centralized world of the double East–West hegemony is fragmenting, politically and culturally, but the homogeneity of capitalism remains as intact and as systematic as ever. The cultural and by implication intellectual fragmentation of the world has undermined any attempt at a single interpretation of the current situation. We have served everything from post-industrialism, late capitalism and post-modernism (as a purely cultural phenomenon expressive of an evolution of Western capitalist society), to more sinister traditionalist representations of the decline of Western civilization, of creeping narcissism, moral decay, etc. For years there has been a rampaging battle among intellectuals[14] concerning the pros and cons of post-modernity, while imperialism theorists have become addicted admirers of all sorts of social movements, and the development elites have shifted interests, from questions of development to those of human rights and democracy. And if the Ferdinand Braudel Center continues to analyze long waves, there has been a growing interest in older civilizations, their rise and fall, and in culture and identity. The intensive practice of identity is the hallmark of the present period. Rushdie's confrontation with fundamentalism highlights the volatile nature of this desperate negotiation of selfhood; the very consumption of modernist literature is suddenly a dangerous act. Global de-centralization is tantamount to cultural renaissance. Liberation and self-determination, hysterical fanaticism and increasing border conflicts, all go hand in hand with an ever-increasing multinationalization of world market products.

(Friedman 1994: 102)

Friedman's global anthropology integrates current political and cultural conditions and applies them to the flow of goods and ideas between societies. His ethnographic work in Greece, Hawaii, Congo and Japan shows patterns of global/local displacements. In Greece foreign images of Greece are 'legitimately' incorporated into present social practices, whereas in Hawaii, foreign images are rejected and only current, local practices are legitimate for Hawaiians. Foreign tourists are a necessary evil in Hawaii, whereas in Japan they are the confirmation of Ainu indigenous authenticity. Finally in Congo, foreign images and foreign goods are used freely to achieve social practices which are purely local (autochthonous). Comparing these patterns, Friedman finds 'the contrast is of symmetrical inversion':

consumption of modernity *versus* production of tradition,
other-centered *versus* self-centered ...

(ibid.: 113)

We now adapt this anthropology of social identity formation[15] and describe the exchanges of knowledge between developers and developees qualitatively. By doing so, we imply the existence of different patterns of developer–developee exchanges which can be isolated and compared, and, furthermore, that these patterns constitute alternative outcomes. Given a certain political, cultural and economic situation in which particular developer–developee exchanges take place, one pattern is always followed. However, given a situation with different conditions, another pattern would emerge. This does not imply that patterns (or habitus) are common to the whole of Chadian society or to the whole of the modern industrial sector of Mexico. Such conclusions can only be drawn when there is a sufficient number of examples of these patterns. Concerning developer–developee exchanges, we can examine only our two case studies.

Friedman introduced the concepts of an exo-sociality[16] and an endo-sociality to describe how an individual interprets foreign goods or ideas to seek cultural confirmation and social identity. He seeks to understand the various phenomena generated by the relation or separation between the essential content and the mode of employment of goods or knowledge moving between societies. These concepts juxtapose the two events I observed. They demonstrate that there are processes larger than any one experience of the passage of knowledge between foreigners and locals, determining the meaning of that passage. These concepts allow us to understand how TA practices are shaped by globalization – as a cultural phenomenon of the present era – a crucial step in observing TA.

In Chad, the passage was an exo-social process because the content of the knowledge did not condition its passage: 'the content does not shape the container' (ibid.: 136). The technical knowledge was used to act upon the cultural distance (alterity) of the actors and to diminish any sociocultural content that it may have accumulated.[17] In Mexico, on the other hand, the conditions of the passage were endo-social. The construction of malinchismo, for example, was a direct result of the Mexican encounter with foreignness. They could not use the content of the knowledge to reinforce their differences as the technical discourse served as an effective bridge between the two sides of the interface. Whatever knowledge came from the outside could be incorporated into the local perspective and vice versa. There was no possibility of attaching 'local' and 'foreign' labels to the knowledge. Engineering of a power-plant in Mexico had international applicability (incorrect even for highly automated plants as organizational behaviourists show).

The exchange dynamics constituted both knowledge and identity for developers and developees. In Appui Technique, the artisans tended to react to the specific behaviours of individual foreigners, and vice versa, even if these differences went unspoken. For example, Martin and Jacques found themselves in agreement with the Chadian experts in their judgement of individual artisans, even though they always avoided discriminating among the artisans. At Autogeneración, the foreigners were frustrated by their failure to reach individual Mexicans through communication. They never found a way of approaching individuals even on a professional level. Significantly, the Mexicans had as much difficulty working with, for example, the US expert of Peruvian origin (Joe) as with the US expert of French origin (Jack). Instead of keeping silent about individual ability, they said it all, even exaggerating at times, but the examples they gave were not accurate enough to elicit a specific reaction.

To summarize then, in Chad the actors believed they understood each other when they communicated, but they betrayed their lack of understanding in the silences that characterized their interaction. Appui Technique exchanges could destroy knowledge when actors expressed themselves; Autogeneración exchanges never got to the knowledge because the actors could not realize their competences. Everything was said in Mexico, but the more they said the less they understood about each other. The content of their exchanges (technical knowledge) was conditioned by their mutual recognition (identity), in a manner specific to each case study.

The empirical differences between these two events are brought to the fore by Friedman's concepts. The end result of the different processes was the same – failed communication. In Autogeneración, the actors were technically close but could not 'see' each other because each actor remained locked in his/her own symbolic imaginary. The encounter was more conditioned from inside; the passage of knowledge had to be negotiated. In Appui Technique, the actors remained distant and could see each other, but failed to allow their lived experience to influence the evolution of the encounter. The encounter was totally conditioned from outside. The passage was not negotiable. The degrees of manoeuvre for the actors in each event were defined predominantly by the specific dynamics. Friedman's theory grew out of the comparison of the endo- and exo-social processes in various cases and contexts. He proposes a global and systematic anthropology that shows how the circulation of goods functions according to the historic possibilities of the constructions of social and ethnic identity in different regions of the world. He understands these possibilities to be localized in a social and historical context. His research agenda has enlarged French contemporary anthropology to include global processes and their role in social identity formation. The two cases exhibit characteristics that invite the application of Friedman's opposition between endo-sociality and exo-sociality to conceptualize the ostensible differences between them and finally locate their similarities.[18] It accounts for the working context of the analytical object called a 'project' based on the lived experience of actors engaged in a transfer of knowledge in two very different environments. Following Friedman, more TA case studies should be used to advance our understanding of these processes.

This first step in defining project dynamics due to exchange patterns interprets actor behaviour at a more aggregate level than the first latent process on technical content. Each actor questioned for him/herself what aspect of the technical knowledge had to be adapted. However, the exchange dynamics between foreigner and local were not accessible to the individual actor. We found the experiences of the local experts to be very different, e.g. José confirming his professionalism and Miguel and Ramón finding theirs constantly denied. Notably, despite these individual differences, all Mexican experts could not separate individual US experts. These dynamics were beyond the actors, latently present in the local context for them to fill in. We describe this latent process further because introducing aggregate concepts can be misleading. Later, mapping out the dimensions of developer–developee encounters (sections 5.4 and 6.1), we will see further evidence of the distinction between exo-social and endo-social processes.

The Chadians looked favourably upon the French presence in West Africa. The Mexicans were less enthusiastic about the US presence in Mexico.[19] The cultural distance projected onto the French experts was thus more easily constructed than that of the US experts. The independent Mexican identity was threatened, notably by the fact that they did not like 'gringos' but wanted their money and so were forced on some level to entertain their way of making it. There were indeed factors other than the colour of their skin that inhibited the clear distinction between these two cultures. The failure on the part of the foreigners to communicate effectively with the Mexicans did not make sense given their vast repertoire of common knowledge and skill. As we have argued, the other was intellectually close and their proximity tended to mask their differences. In other words, the endo-social nature of the interaction in Mexico reinforced the dynamics of the exchanges. Although the defences raised against the foreigners were a function also of the larger sociocultural attitude in Mexico, they were played out through and upon the Mexican experts who could possibly as a group have chosen to challenge the cultural bias (depending on the *habitus* level).

Such resistance was impossible in Chad. The Chadians approved of the

French presence, whose role in Appui Technique represented for them an advancement in political relations. Their defensiveness towards the other was more a function of the radical differences in French culture – a culture whose colour, religion, food, dress and habits were unknown to them – than a protection of self. However, the symbolic domination of the French foreigners drained the technical knowledge that they offered of its substance, reducing the promise of modernization again to a 'white elephant'.[20] For different reasons, Autogeneración saw the same fate. Tragically so, because that project's white elephants were highly appropriate economically and technically. The project dynamics limited the developmental potential of the technological substance.

The juxtaposition of these two dynamics demonstrates that the production of cultural distance (alterity) and identity was intrinsic to the events themselves. If this were not the case in Chad, for example, the cultural distance that marked it from the start would have reduced the need to recreate it, which it did not do.[21] In Mexico, the collaboration should have been facilitated by the pre-existing affinities between the local actors and foreigners, but it was not. The force of the exchange dynamics was determined by the practical execution of the projects. The distinction between endo- and exo-social processes must allow this living dynamic a degree of independence from the context. The passage of knowledge may itself have functioned according to these processes. However, the driving force behind this passage was not the simple need to transmit technical knowledge, but the particularities of each TA encounter. The dynamics due to the actors' identity construction have already been explored in Chapter 3. The exchange of knowledge adds to this powerful dynamic a new, albeit less forceful, layer. Furthermore, the exchanges themselves are embedded in the larger politico-economic history of the countries involved: in this case, Chad–France, Mexico–USA. That is not to say that these historical links are all-determining. On the contrary, they were not, in fact, constitutive of the exchanges in these case studies.

In Mexico, they said that the experts were defensive; in Chad, that they did not get along with each other. But these ostensible relationships masked deeper problems that made social meaning available for the encounters.[22] The primary obstacle to the relationship was the foreigners' difficulty in defining themselves and their role, as well as the contiguous process of definition carried out by the local actors. Deception was not at all a factor in the misunderstanding that plagued the projects. A developer might thrive in an exo-social context and fail in an endo-social one because of the respective social identities. The developmental identity of an actor is not produced by him/her. 'Places' in a global identity space are constituted by practices of 'POSITIONING' oneself, according to Friedman. Developer–developee encounters are fully submitted to such positioning. For the moment, we will not pursue this analysis to its conclusion, this will be easier in section 6.2. But it is important to note the fate of knowledge in these exchanges after having established that the technical content of the knowledge was not the source of confrontation.

It would be unproductive to entertain a discussion of the influence (read: domination) of France on Chad and the USA on Mexico. Unless we could establish the cultural specificity of technical knowledge, i.e. the 'gringoness' of a turbine or the 'Frenchness' of a grain mill, we would have to recognize as an essential quality the Frenchness of the Chadians and the US Americanness of the Mexicans. However, in both cases, these theories certainly ascribe too much power to the foreigner. We would, in effect, force the local actors yet further into the mould of the native in need of development by approaching the question in this way. These concepts stem from the gap between the technical knowledge and the technical mastery of the local experts, which are insufficient grounds on which to base an analysis of the exchanges. The question of whether there are technologies that are non-transferable because they are simply inappropriate is an equally unproductive approach.

Much is to be gained from a specific conceptual approach to exo- and endo-social exchange processes. For the moment, we can point to one qualitative difference. In Appui Technique, the actors used predominantly their preconceptions about the other to explain their work, whereas in Autogeneración the actors used their preconceptions about their own position. In other words, Appui Technique was limited by French ideas about the Chadians and Chadian ideas about the French; whereas Autogeneración was more determined by the Mexicans' thoughts about themselves and the foreigners' ideas about their own conditions.[23] Logically, these are two stages of decolonization.[24] French anthropology has produced an approach for '*la situation coloniale*', and several efforts to update that approach have been pursued.[25]

The exchange dynamics are important for the comparability of the case studies. Since post-colonial history was a very important factor in the realization of these two projects, they were compatible, anthropologically speaking, despite their nearly antithetical socioeconomic historical conditions.[26] Comparing varying case studies is appropriate for isolating new constants for ethnological research. The only element common to the two cases in this study, which shared no sociocultural traits, is, first and foremost, the fact that they were both exercises in TA. Nothing that can be said in a sociological comparison of Chad and Mexico can account for why the fate of the content and identity formation (for developer and developee) was either futile negotiation or non-negotiable. One could speculate that it might be public discourse, political sovereignty or cultural imaginary that creates these outcomes, but, regardless, these constitute specific, singular dynamics.

It follows that these processes at work in the on-site definition of these events are primarily an expression of the endogenous dynamics of the encounters. The articulation of this second latent process depends on the historical and social context. In a particular TA event, it can be more or less prominent, but the same process should appear in all cases in the same context. Nothing occurring between individual actors should be able to change its basic characteristics, despite the causal role of identity formation. The foreign and the local experts, respectively as groups, position themselves by defining their global position as a social identity. Just as the latent process is the result of technical knowledge, exchange dynamics are intrinsic to TA and always remain an essential part of it.

5.3 Interface between developers and developees

So far, we have alluded to an interface (or interactive filter) as the division between locals and foreigners. However, the concept implies more. At the beginning of this chapter, we decided to consolidate the actors' realizations. The most crucial realization is the interpretation of the other side's failure to 'get it'. It is indeed necessary for an individual who participates in such an event to somehow qualify the separation of locals and foreigners. Together, the interpretations by all actors then must approach an equilibrium. The first latent process concerns the content, the second the historical context. This leaves the individual actors, or more precisely their agency, as the subject for the third latent process.[27] We describe the actors' behaviour first in Mexico and then in Chad, and then we contrast the two. The juxtaposition leads us to two types of interface. As before, this is a modest first step to defining an interface in particular events.

The term interface refers to the limit of total understanding within the universe created by the interaction between the foreigners and the local actors as they engage in their collaborative relationship. Long's theory allows us to think about this universe, what he calls a microsocial space, in terms of the communication that takes place between the two groups and the language and/or conceptual barriers that threaten the communication required to work together. The main objective of his theory is to facilitate analysis by observing the interaction between the factors endogenous to the space and other factors such as the colonial past of the two groups. That is to say, interface theory acknowledges the interdependence of micro- and macrophenomena in determining the success of the assistance between foreigners and local actors:

> ... the central importance of treating the researcher himself as an active social agent who struggles to understand the social processes through entering the life worlds of local actors who, in turn, actively shape the researchers' own field strategy thus moulding the contours and outcomes of the research process itself.
>
> (Long and Long 1992: ix)

> ... The concern for interface entails an acute awareness of the ways in which different, possibly conflicting, forms of knowledge intersect and interact. In contrast to Mannheim and Marx, it focuses on the interplay

of different social constructions of 'reality' developed by the various parties to the interface ...

(ibid.: 214)

The concept of interface refers primarily to the fissures created when different cognitive worlds interact and to the means of bridging them.[28] For Long, who is concerned with understanding how unwanted macrophenomena[29] can be generated by conditions internal to microsocial spaces, the interface is articulated in macrosocial structures:

> ... the interactions between actors become oriented around the problem of devising ways of 'bridging,' accommodating to, or struggling against each others' different social and cognitive worlds. Interface analysis aims to elucidate types and sources of discontinuity present in such situations and to characterize the organizational and cultural means of reproducing or transforming them.

(Long 1989: 232)

We locate the articulation of the interface rather in the microphenomena constituted by the personal interaction of the actors. This presupposes that the actors in an exchange can be aware of, and even alter, the interface. Possibly, this is adequate for industrial contexts. The metaphor of the 'arena' that Long has applied is not applicable in this study. The actors were still in a labyrinth and had not yet found the common ground necessary that would constitute an arena.

Analysts tend to use the concept of interface in an explorative way, especially in the social sciences and for public administration. It offers a means of observing the effects of the radical differences between the actors' frames of reference on their ability to work together effectively. What we find is that the actors' capacities to choose between different culturally available frames of reference both increase and limit the actors' importance in the exchange. This capacity increases his/her importance by giving the actor choice, but limits it according to what is culturally available to him/her. In order to employ 'interface' as a heuristic concept, it is necessary to maintain the definition of a social actor, something Long stresses effectively:

> 'Lying at the heart of the concept of social actor, then, is a notion of "agency", which attributes to the individual actor the capacity to process social experience and to devise ways of coping with life, even under the most extreme forms of coercion. Within the limits of existing information, uncertainty and other constraints (e.g. physical, normative and politico-economic), social actors are "knowledgeable" and "capable". They attempt to solve problems, learn how to intervene in the flow of social events around them, and monitor continuously their own actions, observing how others react to their behaviour and taking note of various contingent circumstances.' (Giddens 1984: 1–16).

Anthony Giddens (1984: 9, 14) points out that agency 'refers not to the intentions people have in doing things' – social life is full of different kinds of unintended consequences with varying ramifications – 'but to their capability of doing those things in the first place. Action depends upon the capability of the individuals "to make a difference" to a pre-existing state of affairs or course of events'. This implies that all actors (agents) exercise some kind of 'power', even those in highly subordinated positions; as Giddens (1984: 16) puts it, 'all forms of dependence offer some resources whereby those who are subordinate can influence the activities of their superiors'. And in these ways actively engage (though not always at the level of 'discursive' consciousness) in the construction of their own social worlds, although, as Marx cautions us, the circumstances they encounter are not merely of their own choosing.

Considering the relation between actor and structure, Giddens argues persuasively that the constitution of social structures, which have both a constraining and enabling effect on social behaviour, cannot be comprehended without allowing for human agency. He writes: 'In following the routines of my day-to-day life I help reproduce social institutions that I played no part in bringing into being. They are more than merely the environment of my action since ... they enter constitutively into what it is I do as an agent. Similarly, my actions constitute and reconstitute the institutional conditions of actions of others, just as their actions do mine ... My activities are thus embedded within, and are constitutive elements of, structured properties of institutions stretching well beyond myself in time and space' (Giddens 1987: 11).

This embeddedness of action within institutional structures and processes does not of course imply that behavioural choice is replaced by an unchanging daily routine and repertoire. Indeed actor-oriented analysis assumes that actors are capable (even when their social space is severely restricted) of formulating decisions, acting upon them, and innovating and experimenting. Thus, although one may criticize the premises of decision-making and transactional models (see Alavi 1973, Van Velzen 1973 and Kapferer 1976), social action undeniably entails the notion of choice, however limited, between different courses of action, as well as some way of judging the appropriateness or otherwise of these. Indeed as Giddens points out, 'it is a necessary feature of action that, at any point in time, the agent "could have acted otherwise": either positively in terms of attempted intervention in the process of "events in the world", or negatively in terms of forbearance' (Giddens 1979: 56; for a similar point see Hindness 1986: 115).

Hindness (1986: 117–119) takes the argument one step further by pointing out that the reaching of decisions entails the explicit or implicit use of 'discursive means' in the formulation of objectives and in presenting arguments for the decisions taken. These discursive means or types of discourse vary and are not simply inherent features of the actors themselves: they form part of the differentiated stock of knowledge and resources available to actors of different types. Since social life is never so unitary as to be built upon one single type of discourse, it follows that, however restricted their choices, actors always face some alternative ways of formulating their objectives and deploying specific modes of action.

It is important here to point out that the acknowledgement of alternative discourses used or available to actors challenges, on the one hand, the notion that rationality is an intrinsic property of the individual actors, and on the other, that it simply reflects the actor's structural location in society. All societies contain within them a repertoire of different life styles, cultural forms and rationalities which members utilize in their search for order and meaning, and which they themselves play (wittingly or unwittingly) a part of affirming or transforming. Hence the strategies and cultural constructions employed by individuals do not arise out of the blue but are drawn from a stock of available discourses (verbal and non-verbal) that are to some degree shared with other individuals, contemporaries or predecessors. It is at this point that the individual is, as it were, transmuted metaphorically into the social actor, thus signifying that 'actor' (like a person in a play) is a social construction rather than simply a synonym for the individual person or human being.

(Long 1989: 223–225)

This lengthy citation reveals the actor concept necessary to analyse social agency. Interface analysis suggests, in fact, that the encounter between the foreign developers and the local developees can be understood in terms of the breakdowns in communication and understanding that occur in the attempts to transcend cultural and linguistic differences during the course of the interaction.[30] These moments of failure indicate that, although the actors in the encounter do effectively progress in their understanding of the other in certain situations, they eventually revert back to their previous individual perspectives. In short, each group is unable in the end to grasp fully the nature of its relationship to the other.

With these theoretical references, we define the purpose of the comparison. The schema of the exchange contents is the first tool that permits a comparison between cases. Such a latent process can also be adapted to more specific aims by looking, for example, at the evolution and transformations of each group's perception of the other, or at the argumentation of reasoning used to communicate specialized knowledge. Rather than focus on an isolated instance of interaction, we have conducted a comparison after looking at the symbolic matter of each case (Chapter 4). Comparing the observations made on two different instances of interaction may shed new light on the individual natures of each one. If the interpretation of each event (Chapter 4) could demonstrate satisfactorily the coherence among the different foreign perspectives on the one hand and the reactions of the local actors to the presence of the foreigner on the other, further comparison would not be necessary. The truth of a project lies precisely in the motivation of each group of individuals involved and must be examined in the most practical, and therefore contextual, terms possible. The one-sidedness of each perspective is the fundamental result of this study. The concept of interface will serve as a problematic but necessary framework for the comparison. The first reason for this difficult step is the limited duration of the events. Establishing the coherence of a perspective with descriptive means is difficult for such dynamic and short-lived events.

Moving from the perspective of the foreigners to that of the local actors (in both Chad and Mexico) is a difficult shift. We used the actors' efforts as a guide. The fact that both the foreigners and the local actors in each case expressed general disagreement with the other side's perspective is a solid point of comparison across the exchange. The disagreement helped to determine my position in the event. Nonetheless, their common reception of the observer did not give me an 'objective' view on the difficulties that each side encountered. I also had to extricate myself from my own context[31] while I was participating in order to bear witness to the projects' dynamics, and from the context of each project itself, in order to be able to compare them. The epistemological obstacle to get beyond is specifically the particularity of the more fundamental characteristics of each perspective. In this sense, the comparison between Chad and Mexico does not complicate our task because we face the same epistemological obstacle in both cases: getting beyond. We return to that comparison in section 5.4.

The interface concept does not imply language differences. There was no language barrier between the foreigners and the Mexicans at Autogeneración. If a foreigner said that the team was producing '*bullshit*', a Mexican would know what he meant. Therefore, the foreigner did not say it in the company of Mexicans. In Appui Technique, where Arabic, French and local dialects were spoken, there was a translation problem such that the foreigners could be critical in the presence of the Chadians because they would not understand and vice versa. Crossing the interface is an action of a different nature from translating. The actors use any sort of language differences to alter the interface only because it is convenient, not out of necessity.

After several months of work together and despite the persistence of significant misunderstandings, the different groups could negotiate according to a rationality particular to the project. Although this mode of communication was highly unstable, they managed to make progress and produce some prototypes: a grain mill in the first case and energy analyses in the second. Most of the actors had been working long hours for many months. To maintain the fragile equilibrium of the exchanges, the content of certain signifiers had to be translated, as it were, between developers and developees. It is in this way that an interface could be established between them. In addition, the actors' perception of the other could be transformed in different ways across the interface. An interface does not differ from the other social facts that these actors create in their daily lives. Individuals pursue interests by talking with others. The interface results from two different groups bound together by professional interest. They create it with the means available to them. Unable to define these means explicitly, we regroup their efforts and look at the result, first in Autogeneración and then in Appui Technique. We examine how the interface arose and evolved, and what prevented the actors from grasping it.

5.3.1   Experts bridging the Rio Grande

For Autogeneración, the key transformation which limited the exchanges turned on the interpretation of the foreigners' presentation of their knowledge. They were too, or not sufficiently, authoritarian. The local actors were not conscious of *interpreting* the foreigners' gestures, so often formed a judgement of the foreigner based on their observations. When a foreigner took a pedagogical approach (according to his sensibility of pedagogy), the Mexican expert often considered it to be a kind of deception. Presenting engineering acumen in a simplified manner became, in his eyes, a covert way of keeping him in a state of dependence. When a foreigner presented his know-how in a simplified manner, the local expert was on his guard for fear that it was a trap. If he flaunted his know-how, the foreigner was unable to make it clear and accessible to his Mexican colleague. Across the interface, the effort 'to assist' became an exercise in bad faith.

If this would-be transformation of knowledge was prolonged despite its lack of productivity, it was probably because the stakes involved were sufficiently interesting to both parties. What should have been exchanged there was the career of the expert, the very legitimacy (the power) of the developer. Yet, the foreigners diminished the power of the Mexicans in subtle ways; for example, John suddenly declared out loud '*You know, I am tired of this project. I am leaving tomorrow and I don't care what happens here!*'

His outburst was actually a desperate call for greater effort from his colleagues. Unfortunately, as everyone knew he was going to stay with the project despite his anger, his outburst was futile and served only to belittle the contribution of the Mexican experts. It suggested that his decision to leave was a result of the local situation, which no longer allowed him to put his know-how into practice. No reaction to this accusation was possible and, therefore, no one responded to him. Everyone that heard him continued to work, but with a little less confidence in the recognition of their efforts by the foreigners. The result of his action was, therefore, the opposite of the one he had anticipated and his desperation only grew. John was not capable of understanding that he had not acted in his own best interest.

In Mexico, there is a proverb to the effect that the country tragically finds itself 'very far from God and close to the USA'.[32] Mexico considers itself to be in a position of inferiority in its relationship with the USA, an omnipotent and undesirable neighbour. This sentiment may be easily identified in the attitude of the individual Mexican. When John's sister paid a visit to the project office, and John spoke of an excursion they were planning to make, Miguel commented that the roads in Mexico, constructed as they were by a US company and paid for by the kilometre, were too winding. By making this simple statement, the Mexican expert demonstrated his discomfort with the palpable presence of the USA in Mexico without opening up a debate on the subject. Any kind of debate or confrontation was perceived as a danger, a perilous threat. The actors could protest in the interface but, generally speaking, everyone kept his personal opinion to himself. Thus, they worked together but remained isolated from each other and never engaged in real dialogue.

Without doubt, they were able to appreciate the intellectual contributions of each side. The Mexicans recognized that the foreigners possessed a knowledge that would help them to improve their economy. The foreigners credited the Mexicans with a real professional competency. In a kind of symbolic swap, they each wanted the other culture to validate their professional capacities. In order to please the other, each group had to deny the cultural imaginary of its activity.

John: I think it will be better if I do everything myself – but we always think we could do better.
Ramón: You think you are Superman! [Laughter.] – No, no!
John: Yes, that's the racism of the rich. [One morning in the office.]

The Mexican and the foreigner thus demonstrated their appreciation for the other's cultural mythology: their communal work should have been built on this basis. John and Ramón were the ones who most often had difficulty with the distance between their cultures, and their interactions became increasingly more aggressive. Two months later, John forced Ramón to leave the project, dismissing him as a 'negative element' who had also failed in his work as a manager.

The mutual appreciation should have been an intensive bond since there was nothing standing in the way of their recognition of each other. The experts should have been able to identify themselves symbolically as supreme authorities on the subject, 'private power development'[33] in expert-speak (the Mexican Energy Agency was surely expecting just that), and conceive of their work as driving industrial growth. Why were the actors effectively unable to do this?

Here, we must bear in mind that it would have been imprudent of each side of the interface to concede the legitimacy of the other. In short, to acknowledge before the other that he was able to assert a valid expert judgement was to run the risk of appearing to be less of an expert oneself. The self-consciousness provoked by this fear of disclosure became a defensive wall of silence. Until the end of his participation, Ramón could not admit that he had never worked on the thermal calculations made. Similarly, Carlos could never confide in Jack that he did not possess the necessary competency, as he was, like Ramón, a specialist in microelectronics. In his own mind, Jack rationalized that Carlos's inadequate results were instead the effect of a lack of professionalism. During his interview, Carlos defended his silence as being a means of coping with his work on the project rather than as a means of avoidance:

Carlos: I am not telling you that John is the guilty one, he passes us this; OK, he sees the Agency, he receives the results, and he alone; so if he gives that to us this, not really blaming, I don't think so, he is not such a person to, adequate for managing the personnel, people who are not, like, with attitudes, now I see in a certain way that already this, this doesn't affect me any more, I don't take note of it and I continue to work, and probably Eva[34] can't.

Observer: You filter out what you can understand.

Carlos: But ultimately, Eva can't do that, or I don't know, why there is this, as you say, these jokes that he makes, often comments like, 'I only want to move forward', 'I'll do it with you' and all that I don't care any more. [Interview.]

For his part, Carlos recognized the interface. He was the only one hesitating to accept an interview with me.[35] It seems that it was not his fear of exposing himself as less competent that rendered him unable to demonstrate his expertise or request from the other the recognition of his value, but rather the assumption that the foreign expert would, in any case, be unable to understand him. What constituted professionalism as such was never clearly articulated and that left the actors powerless to rise to the occasion. For Carlos, John's repeated requests for his collaboration were meaningless. He could not see in them an honest and innocent invitation to share his knowledge, so he did not.

The global misunderstanding that led actors such as Carlos to give up on the goals of communication and collaboration was based mainly on the presence of different sets of criteria for the legitimacy of the expert: one for the foreigners and one for the Mexicans. It was evident to me that none of the experts grasped this subtle discrepancy, concerned as they were with more serious misunderstandings generated across the interface.[36] Nevertheless, its presence made itself known in various disagreements and silences that occurred. The most salient evidence of this 'professional' confusion was the actors' incapacity to agree on how to delegate tasks. For the entire duration of the project, the foreigners remained unable to determine in advance what part of the plant analysis could be done by which Mexican expert.

They eventually began to categorize the Mexican experts: this one talks around the problem without precisely identifying it, that one often exaggerates the problem, etc. And yet, it remained impossible for the foreigners to envisage how they would benefit from a Mexican expert's contribution. Despite their good intentions, the foreigners saw all of their attempts to share the tasks fail. The mirror of this interpretation – logically necessary – was that the Mexican experts charged the foreign experts with exactly such an unwillingness to share the work. For example, María maintained that the Mexican experts expected more comments about their own work, and more of a sense of what was adequate and what was insufficient, than the foreigners gave them.

If the interface is the vehicle for the actors to come to negotiate their relationship, it also represented, in Autogeneración, the greatest obstacle to their mutual understanding. The foreigners resolved not to be aggressive or to vex or overburden the Mexicans, but they were frustrated by the feeling that they had been scorned by their local colleagues. The Mexicans were determined not to let themselves be overly influenced by the foreigners, but were frustrated by the feeling that *they* had been scorned by their foreign colleagues. The defensive position of the Mexicans and the rehearsed humility of the foreigners rendered their interaction tedious and superficial. Such a fundamental paradox (intention/effect) illustrates the complexity of the encounter and the power of the interface to affect it.

5.3.2 Approaching and challenging colour

The interface in Appui Technique was stronger, but less visible, than the one in Autogeneración. The cloud of uncertainty that obscured the other side of the interface was so dense that none of the actors could make use of it. Gestures towards the other were more modest. Interpreting Appui Technique,[37] we have already seen that the actors generated three levels of analysis. First, on an explicit level is the purpose of the agricultural machines. Second, on an implicit level, the relation between the teacher–professor and the small-time contractor; and third, on an unperceived level, the relation between the nasarra and the roublard. It seems plausible that these different levels evolved out of the actors' inability to explain what was taking place on the other side of the interface. There were incoherences in what they saw and they remained ambivalent in their behaviour. For this reason, the description of the encounter does bring out these ambivalences.

After the prototypes that they were building were assembled,[38] the experts asked the artisans to make a statement about the project. In response, the artisans assembled independently for the first time (in Osama's house) and drafted a declaration (see Appendix 2) that attempted to clarify their relationship with the experts. They remained, however, unable to incorporate their personal reflections into this document. They had learned the jargon of TA and knew the roles that they were expected to play, but the terms remained foreign to them and failed to capture their experience of the events. The statement did not resolve the first two ambivalences and could not mention the profound one (nasarra–roublard).

In order to have acted on the interface, the actors would have had to express it in one way or another, but they were unable to do so. They could verbalize their cultural prejudices through formulations such as *'as in France'* or *'as in Chad'*, but these statements were too brutal to be useful. Only one other actor, a Chadian expert, was able to act across the interface, albeit with very rudimentary means:

Dambai: What is bad, they are also wary of me because they don't know if I am the antenna, I can always be, um, dangerous.

Observer: Yes, because you can say something to a nasarra.

Dambai: Who would maybe refuse a credit or something. It's a very uncomfortable position, if, effectively, you begin to have a market; it would be very difficult to drain off this market,[39] yes, very difficult. Those that are not even able, he will say, for example, if I slip that to Ngerbo, he will say, 'he corrupted me!' Why do I try to establish myself as, you know, an intermediary? One day Osama tells me, 'but you over there, you are a real antenna!' [Laughter.] I, I sort of grimaced, he was afraid to use this word again. [Interview.[40]]

Dambai saw how the interface functioned and tracked its evolution. One artisan's strategic refusal to say anything to Dambai that could be used against him (as a justification for a credit refusal, for example) was a tactic that the Chadian expert had to take into account. But this strategizing was not the source of the distance between the artisan and the expert. The source was the indirectness of their interaction; a dynamic which, as Dambai points out, would cause one artisan, Ngerbo, to declare that he had been corrupted. If an expert (Dambai, for example) gave instructions to a competent artisan, the latter would not find in the gesture a confirmation of his ability but rather evidence of the expert's corruption. Dambai used all means of non-verbal communication to avoid indirect speech and vagueness. For example, to communicate discontentment, he would grimace, knowing a verbal response would be less effective. During his interview, he explained to me that he sought out occasions for *'lifting the veil'* and for *'putting them at ease'*.[41]

The aggregate of the phenomena that developed across the interface can be reduced to two basic operations. The first involved finding a lexicon of metaphors closer to the concrete, everyday experience of the actors than the 'instructions' idiom of the artisans and foreigners' idiom of 'technological transfer'. Second, these language games were transformed into a simple exchange in which the acceptance of the French developer resulted in the small success of instructing the artisans. The actual exchanges, negotiations of this kind, that were possible in these case studies were limited and involved few actors. The major constraint was the poor visibility of the other across the clouded interface.

It was quite obvious that the foreigners wanted to move beyond a simple relationship of aid and dependence, that they hoped for a purely commercial relationship and that the local actors were eager to dismantle the myth of the 'nasarra' or the *'grand frère'* (big brother). The foreigners frequently promised to establish ties with the local actors as partners rather than as superiors, however they could not overcome the interface. In fact, the foreigners adopted an attitude towards the artisans that forced the latter to play the role of child to their paternal figure. Significantly, most of the artisans were less sensitive to the foreigners' technical discourse and more concerned with the origin of their knowledge. Given the cultural status of the knowledge, they could only respond to the French as apprentices to a master. Other artisans were able to react to the professionalism of the foreigner but this ability remained more or less latent and, according to the testimony of the experts, the artisans often fell back into the 'shrewd' habits. The artisans were thus unable to engage in a meaningful dialogue with the experts. As the following exchange illustrates, the attempts on the part of the artisans to provoke the experts were ignored by Martin, who refused to take part in the symbolic work for or against the image of the nasarra.

Observer: Rahman called you papa again today. What do you think of that?
Martin: Yes. [Laughter.]
Observer: Yes, but I mean, he pushes it a bit.
Martin: Yes, um, yeah, I mean, it's, it's coded language, that is, I mean, I, if you will, ultimately, there is something behind, but I don't even want to know. [Interview.]

The artisans' appreciation of the foreigners' technical knowledge came through in the course of their interviews, but it was never apparent in front of the experts themselves. All of the artisans considered Martin and Pascal, for example, to be very well acquainted with the agricultural machines used in the diverse regions of Sahelian Africa. Confronted with the foreigners, the artisans masked their strong interest in the machines and thus their desire to share the foreigners' knowledge. Even the artisans who felt this interest and desire most keenly could not express it. Across the interface it would have appeared to some as a submission to the foreign experts and the neo-colonial power of their knowledge, and their reservations also held back the others.

Osama, like Martin, was against perpetuating the colonial relationship. Theirs was a relationship stunted by the interface. During their interviews, they made similar assertions against the nasarra image, but finally their appreciation of each other was thwarted by what they labelled inconsistencies in each other's behaviour. Osama's was frustrated by what he characterized as a general incoherence in Martin's discourse, which he further suspected was evidence of a desire to dominate. Martin felt he could not express his appreciation of Osama's work because he thought Osama conducted himself suspiciously. He protested: *'Three times we let him handle the organization and each time they screwed us over!'*[42]

The common understanding of the prototypes, the recognition of the pertinence of the techno-economic analysis (what the projects were really about, after all) and the shared lived experience of the actors were not enough to overcome the barrier that the interface imposed. The other was beyond the interpretive capacities of each side of the divide, whose simple, brutal function was to perpetuate itself.

5.3.3   Where the perspectives were kept apart

As for the latent process caused by technical knowledge, we have reviewed the case studies before inferring common properties. In both, the rare actor succeeded in tapping the power of the interface and pushing through to the other side. Especially John knew how to make use of the interface to support common interest, but Miguel (who often imitated John) also did. Most of the experts (both Mexican and foreigners) at Autogeneración were aware of the interface but could not break it down. At Appui Technique, the interface was stronger still and no one could grasp the transformations of meaning that it produced (with the exception of Dambai). Their efforts to act upon their encounter often targeted similar aspects and their scope of action was severely limited. These limits expressed the developer and developee images and constituted an interface between their interpretations of these roles. To grasp this analytically requires relating the respective rhetorical efforts to these images, abstracting the particular context and identifying how images and efforts fit together. The resulting interface format is latent because it requires constant efforts by the actors. If no actor alluded to its functioning, they would be incapable of working together. Therefore, the properties of an interface can be defined without reducing the ideological transformations that it stabilizes in a particular context.

Figure 5.2 shows the different positions of the interface. In Autogeneración, the interface was visible and the actors could capitalize on the disruption that intervening in the events caused in their reality, even if their efforts proved to be futile in the end. In Appui Technique, the interface existed independently of the actions of the experts and artisans such that the actors could only experience the other indirectly. Consequently, the passage of meaning occurred here between biased images of the other, erected and sustained by the fear and confusion, rather than between the individual actors themselves.

The circle around the foreign rhetoric represents the foreigners' verbal efforts to express themselves, and correspondingly that around the locals. The developee (developer) image is within one rhetorical repertoire when the different aspects of that image are obvious; that was the case in Autogeneración (for instance in the earlier example of Ramón and John alluding to the gringo). The developee (developer) image is partly outside the rhetorical repertoire when parts of this image are too violent to be obvious. Possibly, whenever this is the case, the interface is beyond the verbal efforts of the actors. Since the foreigner never fully understands how a local actor

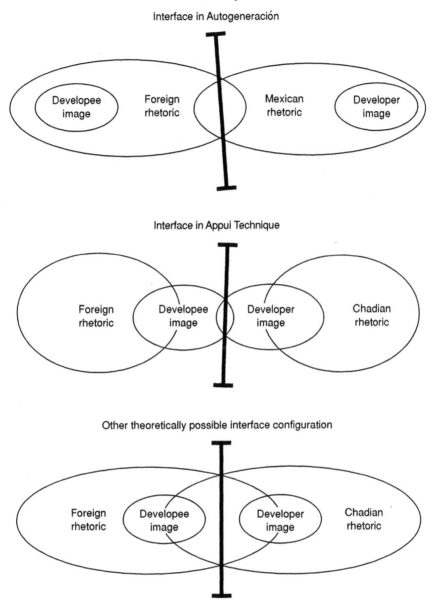

Figure 5.2 Latent process due to the attempts to act upon the encounter

expresses his/her developee self-understanding, and vice versa, the developee image of the foreigner and the local's image of the developer change, align and evolve without the actors realizing how this occurs. These interface formats reflect the structural properties of possible exchanges; 'possible' can only refer to the experts in the two cases. The degree to which most of the developee (developer) images can be verbalized in one TA event is the principal property of an interface. The actors' efforts to shape their encounter is a direct function of this property.

We can speculate about theoretical interface formats that would allow the actors to reach the other's image. In the third case, there should be a strong power differential so that the interface is not resolved by the actors. More interface formats can appear, but we have no basis on which to speculate about them. By identifying and describing an interface, one cannot separate it from the actors, it is not an entity with an independent existence. The advantage of describing an interface is to avoid isolating the symbolism of an encounter. The experts are not bound by the gringo image, the missionary or the colonist. Instead, we infer that the actors reified a specific interface through their encounter. The actors' realizations are defined by the interactionist outcome, the limit of their mutual understanding.

In conclusion, the concept of interface allows us to theorize about the interaction of two groups of people who stumble and fall over conceptual and psychological barriers in their attempts to reach each other across a cultural divide. The actors manage, in time, to adapt to the inconsistencies of the other by ignoring (more or less, according to the actor and the circumstances) the contradictions in their discourse. Such a pseudo-solution does not bridge the gap, but it can establish temporary workable conditions for the experts to function. The exchanges in the two cases thus suggest a latent process. After several years together, the foreigners and local actors on these projects were able to reach each other. They succeeded in developing a system of references that permitted them to act despite the contradictions; and only insofar as their appreciation of the other's value progressed, was received and was reciprocated.

The actors could not take the content of their exchanges to the instrumental core of their knowledge. This made them establish an interface between their perspectives to integrate their interpretation of their roles. Therefore, one reason for this interface is that their professional socialization happened in different cultures. This is a microcondition because it is the actors who produce it. Then, we introduced Friedman's opposition of exo- and endo-social processes to show that the actor's manipulation of the technical knowledge expresses general patterns. This is a macrocondition. All encounters between Chadian developees and French developers should reflect the endo-social process, as suggested by Friedman, and correspondingly in Mexico. The interface connects the macro- and microcondition.

The problem with the interface concept is the functionalist character. Looking at an individual encounter, one can overestimate an actor's power during the exchanges. A perspective of a project is not a private ideology. Long places the actors within their institutional environment, which helps to qualify the actors' repertoire and understand the macroconditions which exclude some options for the actors. Unfortunately, there are no well-enshrined bureaucratic conventions available to the actors that I observed and thus other means are necessary to avoid conclusions that are too functionalist. We attempted to do this by concentrating on those actors who alluded creatively to the differences between the foreign and the local perspective. These creative actions were, in principle, viable for all actors. But they could have changed the basic characteristic of the interface in each case; in Autogeneración the developer and the developee image could not be invested in the exchanges (where the rhetoric overlaps). In Appui Technique, the developer and the developee image did interact, but that was beyond the exchange means. These should be fundamental characteristics of interfaces.

This completes the consolidation of the actors' realizations. We can reconstitute them using the events themselves, disregarding the foreigner and the local actors' institutions, the development agencies, the characteristics of the technologies and the economic interests. This is partly because these aspects were non-problematic to the actors, but fundamentally (actually because it is also a logical consequence) this is possible because there is almost no feedback between planning and TA reality. Three latent processes, concerning the content, the context and the interface, together condition the encounter. Regarding the exchange dynamics and regarding the interface, we abstracted two possible outcomes, although there are many more possible. Each process dissolves a paradox, the first between the confrontation on technology and the agreement over its accuracy; the second between the accuracy and the irrelevance of the experts' products; and the third between the experts' intention and their effects. All three are intrinsic to TA events, latently operated anew by the actors of each event. As such, there is no practical way to stop them and only attempts to modify them towards a theoretical option are feasible – experts seizing the sociocultural relativity of technical ends, experts removing the historical context from their encounter and experts self-reflection on their communicative capacity. We applied social theory to suggest these processes. This yields specific theoretical approaches to be formulated for future empirical verification. Each of these processes requires a book-size conceptualization effort. To stimulate more TA observation, we will overlook this stage for now. While we look for evidence for these processes in evaluation (section 5.5) and management (section 6.3), we continue to add arguments to them.

5.4 Communication structures and the comparative approach

The analysis of my participant observation qualifies the comparison between the cases. The participation in the technical work was not in itself relevant, but the experts reacted to my interviews and took advantage of my presence to act upon the events. My participation as an observer thereby introduced the possibility of comparing their reactions. Functioning as a pawn for the actors, I filled the role that the actors created (section 3.2). Having now pursued this comparison to the end, the result must again be used to define the conditions for comparability. Investigating the verbal efforts of the actors further specifies the degree of comparability between the two cases because the comparison has focused on their interpretation of their exchanges.

The three latent processes are complementary. But they are certainly not contingent on each other because an interface is not necessary, for example, to confound the instrumental core of technical knowledge with sociocultural ends. While each one reflects solid theoretical ground, their pertinence to TA practice remains a matter of judgement. A more precise definition, for example, of an exo-social exchange of knowledge would now have to be verified. Such a definition could then be tested with other empirical examples, but we do not attempt to do so. Instead of embarking on such a verification, we start with empirical examples, and see what tools are available to classify the data. Our data consist of sequences of enunciations. Applying one classification schema, we cannot produce a verification for the three processes, but the classification is complementary to them. The following demonstration of the actors' recourse to a strategic mode of communication points to another difference between Autogeneración exchanges and Appui Technique exchanges. First, we looked at the technical content (Figure 5.1), then at the dynamics of the exchanges, and then at the actors' faculties in all of this (Figure 5.2). Now, in a further abstraction, we highlight their faculties' basic devices.

Possibly, my understanding of the interviews and the notes does not correspond in all aspects to the events. The participants continually and tirelessly negotiated the differences that divided the locals and foreigners. To the extent that I took part in these negotiations, the tension between the two groups of actors was the anchor for the analysis. Writing this text, I placed myself as a foreigner to the foreigner when I described them from the perspective of the local actors and, vice versa, as a foreigner to the locals when describing them. In this chapter, we will clarify this arguably optimistic conception of my position in the field, with the intention finally of problematizing it.

Throughout this analysis, we have reconstructed the emerging logic of the exchanges – a logic always in conflict with itself in its evolution. The breadth of this logic grew wider in the course of the project implementation. In order to make sense of the events, we treated a project as a microsocial space. Even though these projects lasted only several years, the ideological stakes intensified the symbolic exchanges enough to give these spaces the characteristics of a closed world in which the interaction was cut off from its environment. As we have seen, the lived experience of the actors on a project is so complex that the dynamics of the exchanges become an entity that can be isolated and studied.

We have noted the difficulty that the actors had in revealing their common interests in their exchanges and the force of the interface that characterized both events. It is not surprising then to find that they made similar use of the technical concepts upon which the experts constructed their professional identities. Most important, perhaps, were the common subtleties in the structures of their dialogues. The summaries in Chapter 4 demonstrated the privacy of the experience for each participant, but were too factual to allow the structure underlying the exchanges to emerge.

The theory of communicative action exposes illocutionary acts in a dialogue which allow interlocutors to build up a horizon of shared meaning.[43] If the interlocutors maintain without reserve an illocutionary target during one or more dialogues, they establish a co-ordination sufficient for the creation of new meanings and interpretations in terms of exterior events, but especially in terms of their mutual rapport. The illocutionary part of the utterance is defined as that element which orients the content of the statement towards a particular goal. If the speakers involve new interpretations, their communication becomes stronger and their language acquires a social energy that creates more intensive rapports. Without going into the important implications of this theory for social science in general, we can examine the concept of illocutionary acts to see if it will enable us to advance, notably in terms of the distinction of communicative, strategic and dramaturgical acts.[44]

The misunderstandings across the interface discussed earlier reveal the efforts made by the experts to achieve a satisfying co-ordination and to generate new meaning in Autogeneración. The following excerpt from a conversation between Eva[45] and John is a good example of a dialogue that fails to sustain an exchange of meaning:

John:	You know what?
Eva:	No.
John:	You won. We're going to do everything in Courier font.
Eva:	Aha.
John:	Yes, it's your Mexican eye.
Eva (turning to Juan):	You see, John is a racist.
John:	No. It's a compliment. Is there no pride here?
Eva:	Yes, a lot!

With the verb 'to win', John demonstrated that his utterance should perform the act of ceding victory to Eva. In the absence of a response which confirmed that his target was confirmed, he qualified his statement differently in a form closer to the illocutionary targets that Eva had proposed earlier (a Mexican eye). Eva ignored John's acknowledgement of the targets she had proposed and, addressing herself to a Mexican expert (Juan), she asserted that John was categorically incapable of following her. It goes without saying that the decision to use Courier as a font had nothing to do with this exchange. Both were aware of this, and they knew, too, that John was not a racist and that Eva was not a clairvoyant. John continued to affirm the target of his argumentation (to cede victory) by making it more explicit. In order to contest John's claim that his illocutionary act corresponded to their reality, Eva countered that John's intentions were misguided from the beginning of the exchange.

Many of the exchanges among the experts unravelled in this way and led nowhere. The use of illocutionary targets was often a unique event. Each expert tested his/her limits by appealing to one part of the engineering work, or to a colleague. The experience of a failure made it difficult to repeat the attempt because any such failure reminded the actors of the failure to collaborate; and thus the loss of the project's *'raison d'être'*. In response to the small failures, the actors had two options – two ways of continuing their exchanges over the long term. They could treat their utterances in the exchanges as acts of dramaturgy, or as strategies:

> As Habermas puts it, there exists no logical continuum between the various types of utterances. Thus each function represents a pure case of a type of speech act. The illocutionary component gets to the core of the social bond created in linguistic communication: its analysis captures the generative function of speech, which on the structural level integrates speaking and acting through the medium of intersubjectively shared meanings and binding validity claims.
>
> (Bohman 1986: 336)

> I speak either of 'communicative action' or of 'strategic action', depending upon whether the actions of different actors are coordinated by way of 'reaching an understanding' or 'exerting influence' respectively. From the perspective of participants, these two mechanisms and their corresponding types of action mutually exclude one another. Processes of reaching an understanding cannot be undertaken with the dual intention of both reaching an agreement about something with a participant in interaction and having an effect on him. From the perspective of the participants, an agreement cannot be imposed from without, cannot be foisted by one side upon the other – whether instrumentally, through direct intervention into the action situation, or strategically, through indirect influence, again oriented only towards success, on the propositional attitudes of the other. Whatever manifestly comes to be through external influence (gratification or threat, suggestion or deception) cannot count intersubjectively as an agreement; an intervention of this sort forfeits its effectiveness for coordinating action.
>
> Communicative or strategic action is required when an actor can only carry out his plans of action interactively, i.e. with the help of the actions of another actor (or their omission).
>
> (Habermas 1992: 79)

> From the vantage point of communication theory, by contrast, strategic interactions can occur only within the horizon of lifeworlds already constituted elsewhere, more precisely, as an alternative option in case of the failure of communicative actions.
>
> (Habermas 1998: 248)

In dramaturgical action the actor, in presenting a view of himself, has to behave toward his own subjective world. I have defined this as the totality of subjective experiences to which the actor has, in relation to others, a privileged access ... In the case of dramaturgical action the relation between actor and world is also open to objective appraisal. As the actor is oriented to his own subjective world in the presence of his public, there can be *one* direction of fit: In regard to a self-presentation, there is the question whether at the proper moment the actor is expressing the experience he has, whether he *means* what he *says*, or whether he is merely feigning the experiences he expresses. So long as we are dealing here with beliefs or intentions, that is, with cognitive acts, the question of whether someone says what he means is clearly a question of truthfulness or sincerity. With desires and feelings this is not always the case. In situations in which accuracy of expression is important, it is sometimes difficult to separate questions of sincerity from those of authenticity. Often we lack the words to say what we feel; and this in turn places the feelings themselves in a questionable light. ... dramaturgical action can take on latently strategic qualities to the degree that the actor treats his audience as *opponents* rather than as a public. The scale of self-presentation ranges from sincere communication of one's own intentions, desires, moods, etc., to cynical management of the impressions the actors arouses in others.

(Habermas 1984: 93)

We look first at examples of dramaturgic acts and then at strategic acts in Autogeneración in order to qualify the shift from communicative action to dramaturgic action and then the shift from communicative to strategic action. Undoubtedly, these two shifts were only a small part of the actors' communication competence, but these shifts are particularly suitable for characterizing these exchanges. It is not productive here to analyse why an individual actor chose between dramaturgical and strategic action. By excluding this question, we assume that these choices reflected only individual habits rather then a general condition. Furthermore, the failure of illocutionary efforts is too general a condition to be of interest, intercultural exchanges are always difficult.

Dramaturgical acts are essential for the construction of identity. According to George Herbert Mead,[46] only dramaturgical acts allow an actor to show and reveal to him/herself his/her relationship to social norms. Dramaturgy thus becomes even more important in intercultural communication because the inevitable violation of social norms that accompanies contact creates the need to affirm more often that the actors are acting morally. In intercultural situations, the dramaturgical acts are, as it were, the pleasure and fruit of the work, of the difficult communicative acts – the harsh realism of the everyday. The pleasure in dramaturgical exchanges reflects the scarcity of opportunities to prove one's morality, it comes as a relief. Illocutionary targets fail so often and require such constant concentration and attention that the actors do not find much comfort when they make progress. Within one's own society, the opposite is true; we feel at home where we act communicatively with ease and social relations are rich.

According to Habermas's theory, dramaturgical acts do not permit the subject to induce new meaning in the dialogue. However, these acts may well serve as a means of continuing the dialogue when there is no shared lived experience to which one can refer. The following exchange between John and Miguel illustrates this usage of dramaturgical acts:

John: Are the cheques signed?
Miguel: Don't ask me because I'm tired of this subject.
John: Why? Wasn't Isabel there?
Miguel: No. The cheques were all in my office, locked.
John: But how can I take this if I have to stay here for 6 more months?
Miguel: Well, by retaining your sang-froid!

Each interlocutor dramatically revealed the perspective through which he received and interpreted the others' acts. This process invited a return to the attempts to harmonize the various perspectives on the project. To continue working together, even at a distance, they avoided dramaturgical acts and thus reverted to strategic acts.

The Mexican experts saw many of the exchanges as latent strategic acts. By rejecting an illocutionary target, they protected themselves from being implicated in the suspected strategy of the foreigners. For example, the promises to allocate the outstanding engineering tasks lacked meaning, knowing that the other would not fulfil his part. When actors frequently use the strategic mode, the interface grows stronger. Each side assumes that it cannot penetrate past the interface to the other and the logic of the other side remains hidden. Finally, it is difficult to know whether this breakdown is the result of one side wishing to hide itself from the other or whether it is the result of a genuine incapacity on the part of each side to understand the other.

Once the exchanges appeared to be purely strategic, the experts on each side quickly adapted and responded to the other's strategy and the exchanges degenerated badly. The simplicity of the exchanges between the foreigners and the Mexicans is reflected in the following example of how the foreigners prepared for them. John suggested to Bill that he should make a visit to a plant with a Mexican expert:

John: It's a three day trip.
Tom: I can't, I've got interviews set up.
John: Bill, do you want to [go]?
Tom: Oh, yes. He can verify the data while the rest of us talk.
Bill: Yes, let them bullshit while I do the work.

Tom:	Yes, and it won't even be offensive because, in any case, you don't speak Spanish so you can't participate in the discussion.
Bill (smiling):	OK, we'll see if a dumb gringo will help the flow of information.
John:	Well, Tom went to the chemical plant and it didn't help.
Bill:	Yes, but he's not a dumb gringo. [Discussion during a lunch break.]

The experts resigned themselves to orienting their actions in terms of the non-communicability of their know-how. Neither the supposed beneficiaries in the plants nor the Mexican experts could use it. Bill's language barrier could not be counted among his limiting factors because even the bilingual foreigners failed to communicate effectively.

To confirm know-how and identity as José was able to, it was necessary to overcome the strategic mode of exchange. If his success had been shared by his colleagues, other Mexican experts would have been able to do likewise. José's superior technical competence permitted him to go further than the others in proposing and recognizing illocutionary targets and in seizing opportunities to operate outside the strategic mode in exchanges with the foreigners.

John and Ramón, two young and eager experts, proposed risky targets when they uttered '*racism of the rich*' and '*Superman*'. John often directed the conversation ('*we always think we can do better*') and turned the screw by exposing his own limits. Ramón was able to show that he did not liken John to Superman but this symbolic work was not enough. He eventually left the project because he could not live up to his own image of Superman. The efforts by John and Ramón to identify themselves according to the other's metaphor was not an atypical strategy. The actors had to define their encounter intersubjectively – the only means of achieving identity – because within the project they were strangers to the social meaning of the other. This intersubjective definition was particularly minimal because, in general, the actors distrusted their own ability to make the other recognize the claim to validity that characterized their utterances.[47] The establishment of this essential confidence was impossible notably because of the limited duration of the project. Its impossibility was in a more general way also caused by the lack of popular notions of TA. There was no acknowledgement of a Mexican expert who made use of foreign knowledge, nor of the foreigner who presented technical knowledge without reference to its origin.

The examples from Autogeneración show the rhetoric of endo-social exchange processes. They show how the interface created the conditions for these exchanges. Further, they demonstrate that dramaturgic acts were rare because the actors managed with the strategic mode. This is another indication that constructions of identity and cultural distance (alterity) are authentic to these events. They drive the exchanges. The theory of communicative action helps to identify the possible directions that an encounter can take depending on the nature of the exchanges. We can observe the effects of an exo-social context in the exchanges in Appui Technique.

Over the last few years, in the aftermath of the project in Chad, it has become clear that the discussions that occurred during the experience did not generate lasting ties. Every actor might have asked himself 'did I get it wrong because I did not understand what the other wanted, or did I hesitate too often to explain myself?' Certainly they interpreted boldly and often thought they understood each other despite the deep-rooted mis-understanding. And, occasionally, a genuine moment of complicity did occur. Dambai and I were able to identify the seed of our distance in our mutual appreciation of one of the foreign experts. Dambai thought that criticism was impossible, while I considered it necessary. Our exchanges passed rapidly from one illocutionary target to another, and the difference was explicit. This difference was not a moment of failure but merely a juncture where Dambai reached the limits of possible reasoning – limits imposed upon him by the context of his circumstances. In our exchange, the differences between the foreigner and the local actor were visible and could be articulated. The limits to our agreement were normative constraints imposed through the interface from the outside.

The gesture of communicative acts in order to reach the other was still possible, and, often, when an actor endeavoured to be understood she/he accomplished that. However, actors rarely sought out and seized possible opportunities because each one worked within cultural limits that were imposed to avoid them. Osama affirmed that critique of the foreigner seemed impossible:

Osama: No, no, no! You can't tell him that – if you tell him that, no, no! [Interview.]

Nonetheless, the active courting of the other was a risk on both sides. The local actors, in particular, protected themselves from the other who was stronger than they were.[48] Communication may have been mechanically functional (a dialogue did exist), but because of the cultural distance perceived the symbolic work necessary to real understanding was a monumental task.

The distance separating the foreign and local actors rendered dramaturgical acts impossible for the most part, but there were notable exceptions to this rule that involved my presence. Mohammad dramatized my position, but he and Osama were the only actors capable of doing so. During the final dinner to celebrate the completion of the oxcarts and the grain mills, Jacques and Dambai addressed the general public, but the artisans did not respond. Rahman attempted to provoke a reaction but failed. During a final meeting the next day, Jacques and Mohammad again used dramaturgy; I was due to leave N'Djaména 2 days later:

Atula: It's a shame, but we can't change the plans.
Jacques: We must marry Thomas off here!
Mohammad: He could easily do it. We wanted to give him a Chadian woman, but he refused.

Jacques:	Ah!
Dambai:	He already tried.
Mohammad:	He says that Chadian women want children but he can't take them home with him.
Pascal:	Have you already met the woman's father?
Mohammad:	But if the woman agrees, there's no problem with the father.
Pascal:	And you already arranged it?
Mohammad:	No, he refused! I can't do it!

Mohammad seized an opportunity to express his desire to make me part of his Chadian world. Jacques, who had made his remark in jest, was amazed that others should take it seriously. What he assumed to be an obviously humorous suggestion registered as a real proposition for Mohammad.[49] Dambai's comment further supported Mohammad's claim that he wanted to assimilate me into his culture. Jacques, Mohammad and Dambai all tried to perform to a certain extent, but they did not take advantage of the possibility of using these dramaturgical acts to strengthen their impact because it would have made the discussion too hazardous. The general role of the observer was particularly prominent in giving the actors opportunities for dramaturgical acts (the example of Mohammad, pp. 40–41, is quite typical). My passivity during my exchanges with the actors reduced the risk of further deterioration of the relations.

In the endo-social situation at Autogeneración, dramaturgical acts were possible. Owing to the exo-social conditions at Appui Technique, the actors could not invite the other to engage in symbolic work. The actors' acceptance of the strategic mode in Autogeneración signified defeat for them. In Appui Technique, on the contrary, this was not lived negatively and considered to be a failure, but was accepted as inevitable. This qualitative difference is the pragmatic analogue of the difference between endo- and exo-social processes. The limits of their comprehension of each other were less visible in Appui Technique than in Autogeneración. The pragmatic analogue allows us to see another element of the difference between exo- and endo-social exchange processes and to see whether these two configurations are alternatives.

The strategic mode of exchange between the foreigners and the Mexicans enabled them to sever their ties after the project ended. The artisans and experts of Appui Technique followed up their efforts on the project differently. The factors limiting the exchanges in this case were imposed from outside; the strategic communication seemed to them not to be generated by their encounter. Consequently, this imposed condition had not become an obstacle between the foreign and Chadian experts; Dambai, Atula, Tahem, Jacques, Martin and Pascal were all able to discuss their relations with the artisans quite explicitly and the foreigners used dramaturgical acts to explain themselves to their Chadian colleagues. Each of them found ways of explaining his role and making sense of the exchanges. We can speculate that the presence of the artisans and the difficulty of relating to them brought the experts closer together, and enabled them to overcome certain limits by projecting conflict onto the artisans. By comparing themselves with the artisans, the experts recognized their own similarities, a recognition that can be characterized here as passive. Out of this forced company came a strange sort of solidarity. This was explained to me by:

Dambai: You are conscious, all the same, of certain things, of all this shit that we have every day, the Chadian knows that there is the one, the one people take for a savage – yes, it's true, we should not be afraid of words – whether he's nasarra or black, that lifts the veil on, um, there it is, but I don't have the right to go tell Martin: 'It's an insult to say that …'. [Interview.]

Dambai felt he did not have the right to explain to Martin why some artisans resented Martin's behaviour because he knew that Martin did not mean to cause any offence. Had Martin spoken to the artisans as he did to Dambai, the artisans would have reacted in a similar way to Dambai.

If communication is functional but not effective, it is far less efficient as a co-ordinator. Dramaturgical acts include performative attitudes that cannot be intersubjective, usually because of a lack of common ground, and that suffer from the absence of codified social norms. In front of their public (the rest of the participants), the actors' subjectivities essentially remained hidden. A long time before the experts realized that the collaboration with the artisans was going to fail, the collaboration had ceased to work because of the accumulation of misunderstandings. Their lived experience could not be objectified with practical knowledge.

By revisiting the comparison between the case studies, we have demonstrated the limits of an interactionist analysis, an approach that conceptualizes exchanges without recourse to the lived experience of the actors. It is vital to follow the action of the participants, and this knowledge shapes our understanding of the foreign–local actor relationship. Thus, we return again to the exo-/endo-sociality distinction. We have not been able to improve on this, and conclude that the participants, who did not recognize their exchanges as dramaturgical or strategic, were not consciously limited by these defences. The position of the other *vis-à-vis* the interface was a more fundamentally defining element for the exchanges.

The necessary conditions for the juxtaposition and the comparative analysis of the case studies have also been clarified in this methodological section. Having recourse to strategic exchanges was negative in Autogeneración (a defeat) and neutral in Appui Technique (an inevitability), but, in both cases, the acceptance of non-communicative exchanges characterized the interactions in general. Having begun with the lived experience of the actors, this defining element of both life-worlds constitutes the bridge that links them as a point of comparison. We were able to join these projects without reference to the various differences that render a comparative study difficult:

their very different histories, the various nationalities involved and the spectres of their colonial or imperial past.

In other words, the common element that can be detected in these two projects is the passage of knowledge from one culture to a very different culture through the communicative collaboration of their actors. There is a familiarity among even the most divergent case studies of development experiences (in their failure) which implies that language games offer a way of defining the logic of the communication that is specific to a developer–developee encounter. With the help of Friedman's concepts, we have concluded that the passage of knowledge follows the rules of either an endo- or an exo-social process. The subsequent differences, born of the interaction within the space of exchange, can be read as a function of this fundamental condition of the transfer of knowledge.

This conclusion is the consequence of a laborious abstraction, but it serves three purposes. It shows to what extent the individuals described are 'actors' in the sense that there are limits to their means to act. Then it illustrates how macroconditions translate into microexpressions. Thereby, it reinforces the types of exchange dynamics suggested. And, finally, it helps to qualify the comparison between Autogeneración and Appui Technique. Comparing them does not confuse colonial domination in Latin America with that in Africa, the nasarra with the gringo, or 'high' and 'low' technology. From a pragmatist's point of view, comparing Autogeneración exchanges with Appui Technique exchanges is like comparing TA where the strategic exchanges are a personal defeat with TA where strategic exchanges are the 'normal' state of affairs for everybody.

5.5 Evaluation science

We now reinsert all of the results noted so far into TA practice in general. How do the results compare with those that are recorded by the development agencies? Within these agencies, observations about TA are called 'evaluations'.[50] The scientific field of evaluation is well established, with textbooks, professional associations, journals and a market for its services, and is a part of the development industry. Before reviewing it, we will proceed with a thought experiment, a projection based on the latent processes. The choice of representative evaluation documents is difficult, but it serves to illustrate the projection. This serves to validate the latent processes as well as to clarify what role evaluations have, a task that many development agencies struggle with. Finally, we use the case studies to point once again to the sheer insurmountable conceptual problems for meaningful evaluations.

The number and volume of monitoring reports; quarterly, biannual and annual project reports; special evaluations; terminal reports; debriefing reports; and appraisals is breathtaking. Kim Forss estimates that around 200,000 evaluation documents have been produced by agencies in the past. Every development agency has its own evaluation procedures, with unique jargon, data and variables that are, in general, difficult to compare. Because of the closed character of the insider perspective, evaluations of any type are limited, but their effectiveness still compares well with TA projects themselves. As one would predict, some development agencies spend considerable resources on evaluations.

The actors of both case studies read this text approvingly, drawing conclusions which reflected the development agencies' response to our results, while being unable to overcome the contradictory interpretations they had built up as a result of the interface.[51] The insider perspective of TA is fraught with contradictions aroused by the encounter between foreign and local actors. The recognition of technological content, the dynamics deriving from the exchanges of knowledge and the interface allowed us to reconstruct the outcome of these encounters. This explosion of ideology disappears along with the actors at the end. Attempting to reconstruct the events from the outside, one is left with many open questions and without the means to answer them. The traces left in the actors' memories, if they dare to express them, make little sense. An outside evaluation is indeed an almost impossible task. Calls to abandon evaluations pointing to the complexity of implementation date back to the 1970s.

Without a doubt, none of the three latent processes appears explicitly in evaluations. On occasion, allusions to communicative competence appear. First, because evaluations are elaborated from the outside, and, second, because the response of development agencies cannot contradict official development discourses. Consequently, a direct comparison between the latent processes and evaluations is not possible. However, it is feasible that these processes may be treated as hypotheses and, assuming incorrectly that our case studies are representative of many TA encounters, that we can determine what the outside evaluations yield. This projection requires one assumption: that the number of evaluations is so large, and the methodologies were redesigned so many times, that the open questions from outside observations have created hazardous conclusions that reflect the needs of development agencies. This assumption is correct in many sectors of TA. In other words, evaluations have evolved into an element of the reproduction of TA encounters, and the weakness of evaluations reinforces the underlying causes for TA success/failure. This is vehemently denied by evaluators, but those who are critical of their own evaluation experience point to the key role of the institutional response, determined by political needs, e.g. Partridge in OED (1995a: 20) or Carlsson *et al.* (1994: 180–183). We now proceed with the projection, in the order of the latent processes.

What happens if TA always enacts the content process presented in Figure 5.1? The actors could not seize the symbolic overdetermination of technology nor distinguish instrumental from sociocultural ends. Within one society, this is analytically useless because of the seamless web between culture and technology.[52] In TA between societies, this is not only possible in principle but also happens more or less fortuitously. Foreign and local experts act on each other's imagined competences and thus cannot seize the instrumental core. Outside evaluations can conclude that foreign and local experts share the instrumental core by analysing these experts' competences individually after the end of the project. The evaluation result should be that, whatever the knowledge concerned, even if the locals have it and the foreigners have it also, they seem not to share it when they are working with each other. Concluding this at least initially, evaluation science should realise that expert competence correlates only negatively with TA project results, i.e. only the absence of competence is a factor, negatively, but there appears to be no corresponding positive correlation that higher competence would bring improvements. This would reflect the content process.

We then suggested that the exo-social exchange dynamics do not allow the actor's experience to influence his/her encounter, and the technical content could be dissolved in order to create cultural differences (alterity) between the actors. The difference from endo-social dynamics should be quite marked during evaluations; possibly to such an extent that there are methodological specificities. Since exo-social exchange processes always occur in a particular segment of TA (region, industrial sector), evaluations have become adapted to these dynamics, and correspondingly to other types of exchange dynamics. Let us suggest, for the sake of argument, that there are specific dynamics, typical of the informal sector in sub-Saharan Africa. When technology can be dissolved to create cultural distance, evaluations should lead to the conclusion that technology transfer is categorically impossible. Local technology and technical experts become enigmatic. They can only advance by their own efforts, if they so wish. Indigenous knowledge will be called for, even where the concept is difficult because there is no locally available experience. The 'informal sector' concept itself reflects this, it is only negatively defined – it is not formal. Equally, actors' implementation skills should appear to be undefinable. They seem not to acquire any skills besides getting used to ignoring their particular experiences (as was the case with Chadians and French).

In endo-social exchange processes, the actors' competences appear, but they cannot apply their competences to their encounter; this is all the more frustrating for them because they feel that they are the source of the 'misunderstandings'. Without other case studies, one can speculate that this endo-social dynamics appears in large infrastructure-related projects (power, transport, telecommunications) in Latin America. Evaluations should then lead to the conclusion that there are extraordinary rigidities in organizations. The experts recycle the technology which works elsewhere, but they cannot reap the benefits. Technology transfer seems feasible but somehow creates its own obstacles. And, strangely, the obstacles visible from outside one encounter are not the same as those visible from outside another encounter. This shows that the obstacles are always the product of the particular actors. Consequently, there is no reason why different evaluations should lead to any coherent aspect between them. The pioneer observer of development projects, A. O. Hirschman, coined a term for this: 'fracasomania' (fracaso is Spanish for failure).[53] Failure appears to be the result of passion or obsession because from outside the encounters evaluations cannot isolate any determinant of TA results.

Finally, what are the consequences of the interface process (Figure 5.2) for evaluation science? Again, there is nothing to suggest two structural formats of the interfaces. Possibly, there are many such formats and hybrid versions. Nonetheless, assuming that there are interfaces between interacting developer and developee images, even though these images are beyond the rhetoric of the actors, and assuming that there are interfaces where the actors allude to the interface and thereby isolate their developer (developee respectively) image from the interface, evaluations are based on different kinds of explanations from these actors. Where the images interact outside the rhetoric, the actors perceive betrayal or loyalty, and developers and developees judge one another in terms of normative conditions (the Appui Technique – type interfaces). Alternatively, when the actors use their imagination (Superman, racism of the rich, etc.), their recollection should appear in terms of degrees of expert worthiness, professionalism, intelligence and so on. In the latter case, there are those who live up to their 'mission' and those who are categorically not able to. Foreigners and locals accuse each other of lacking professionalism (as did the US and the Mexican experts) despite their explicit and mutual recognition of skills. If this juxtaposition is appropriate, evaluations should show that both foreign and local actors make either moral statements or competence statements. Furthermore, this should be uniform, irrespective of whether the individual asked was the driver, the telephonist, the leader of the party or the chief scientist. Two types of genuine 'implementation topics' should appear: morality or professionalism. Evaluation reports, if they manage to gather the actors' impressions, should not record a mixture of both. Table 5.1 resumes the thought experiment, summing up the projection for each process. This table should be intuitively convincing to experienced producers of evaluations.

The projections are plausible, but, to complete this thought experiment, the hypothetical conclusions must be compared with the actual ones in development agencies' documents. This is difficult because the grey literature on TA evaluation is huge, difficult to access and inconsistent. To comment on the thought experiment, it is necessary to choose some particularly pertinent evaluations.[54] We review the best European, US and IBRD TA evaluations.

A particularly comprehensive example is an evaluation of Scandinavian TA produced by Kim Forss and co-workers (Forss *et al.* 1988).[55] TA is well developed in Scandinavian agencies and the evaluations are comparatively sophisticated and thorough.[56] The study by Forss *et al.* (1988) is widely cited and has contributed to a reduction in long-term assignments for foreign experts. It concerns four different development agencies, and it enjoyed many means of questioning experts before, during and after TA engagements. More than 800 interviews were conducted about fifty-five projects in Kenya,

Table 5.1 Hypothetical evaluation conclusions

Condition	Conclusion appearing in evaluations when observing TA practice from outside
Impossible to separate instrumental and sociocultural ends of technology (Figure 5.1)	Expert competence is weakly correlated to better results, only insufficient competence is a strong (negative) factor
Exo-social exchange dynamics	TA appears to be impossible, frequent calls for indigenous knowledge, no implementation skills identifiable
Endo-social exchange dynamics	TA appears to be possible, but creates its own obstacles, fracasomania
Interface (Figure 5.2)	Locals and foreigners make all normative accusations, or all see the others' lack of professionalism, a mixture of normative and professional comments never appears

Tanzania and Zambia, which were funded by SIDA, DANIDA, NORAD and FINNIDA. The study produced many operative recommendations for enhancing the effectiveness of technical assistance personnel (TAP). Forss *et al*. (1988) attempted to improve upon the work carried out in Canada in the 1970s by Hawes (1979), whose reliability and validity they doubt; 'in consequence we will not take for granted that any personal characteristics predicting success really exist' (Forss *et al*. 1988: 8). The following quotations are most relevant to the recommendations because they express the opinions of the local and foreign actors:

> Development projects are risky by nature. They are often attempts to start new activities or to expand activities into new environments. The only method existing is that of 'trial and error' and success ratios of more than 30% must be considered very satisfactory.
>
> (Forss *et al*. 1988: 32)

> ... we must conclude that the aim of the TAP has not been fulfilled to a satisfactory degree. But we have also seen that the situation is not static and that elements of existing TAP projects and new donor procedures are pointing forward.
>
> (ibid.: 43)

> ... the counterpart[57] situation is a problem in around two third of the projects, and about half of the TAP, who were supposed to have counterparts, did not have any.
>   One conclusion is that the old combination of on-the-job training and counterparts is not working, and not solving the training problems. In

the following some of the reasons are given. First as seen by the recipients:

It has been donor thinking, that professionals in the recipient system get new possibilities in their personal development and new career opportunities by being counterparts at a project with funding and technical assistance from outside.

But the reality is much more complex. As expressed by a provincial permanent secretary in Zambia: 'Counterpart positions are not popular or attractive. They are subordinate and low paid, and it is often very unclear, which career possibilities they give'.

This statement has, with variations, been repeated in all three countries. In some ministries and sectors it seems obvious, that national authorities are not ready to post their best people as counterparts. Counterpart positions can be dead ends for more senior people.

Very many of the counterparts interviewed have criticised their positions for being unclear. They do not know if or when they are going to take over responsibility or how long they have to be in a subordinate position.

(ibid.: 50)

TAP at projects will often see the counterpart problem as a proof of lack of priority given to the projects by national authorities. They are not given the counterparts they have been promised, or they are given too young inexperienced people. Sometimes they have to accept that good counterparts are transferred to other positions.

One of the conclusions is, <u>that counterpart systems are seldom effective. A minimum condition is that such systems are parts of more formal and systematic manpower development plans included in project plans from the beginning, based on time planning, career development plans for nationals working as counterparts or studying and on more systematic training plans.</u>

(ibid.: 52)

So far, Forss *et al.* have concluded that transfer of knowledge rarely happens. Therefore, they recorded data about the general conditions of local experts' employment, career prospects, etc., which could prevent the local experts from benefiting from the TA encounters. This was an odd thing to do: if counterparts do not acquire skills or only irrelevant ones, why offer other benefits?

One way of measuring discrepancies between the original aim of the TAP component and the factual situation today, is to look at the number of projects, which have had both a longer lasting and a bigger TAP component than originally anticipated. ... In more than half of the evaluated Tanzania projects, and in many of those with the biggest TAP component, the development of the TAP situation over the years had very little resemblance to the anticipation of the planners.

(ibid.: 59)

Sometimes a vicious circle has developed. Difficulties in recruiting and retaining local manpower and lack of local involvement leads to additional TAP posts. Additional TAP posts lead to an expatriate community less integrated and still more with its own social life and norms, but with the ability to manage and implement all project elements. This situation leads to deliberate (but seldom open) decisions giving less priority to the national or local manpower support to such projects. In a situation where the recipient has a manpower scarcity and lacks funds, local top priority will seldom be given to support for projects, where the whole management and implementation formally or in reality are in the hands of the donors.

(ibid.: 63)

Host country attitudes to TAP:

It is generally accepted that TAP are very efficient in the implementation of their projects. The reason for this is seen to be the good facilities they have, not their personal or professional qualities. Many local officers claim that they could achieve the same, given the access to the same equipment and other facilities.

The locals are especially critical of the fact that they are not given an equal access to the project vehicles with TAP. They are aware that they are suspected of misusing the vehicles but, as they see it, also TAP misuse the vehicles by using them for private driving and holiday trips. Generally it was stated that misuse of equipment and dishonesty are not confined only to local personnel but exist also among TAP.

(ibid.: 77)

... The counterpart system has not given expected results. Cases in which a counterpart has actually succeeded a TAP are very rare. Counterparts are often frustrated because when one TAP leaves another one comes in her/his place. Being a counterpart is not an attractive position to a local officer because they see it as inferior to a position where they are fully in charge. A counterpart is also often away from the line positions in the ministry and consequently misses promotions.

The most common problems experienced by TAP:

The most common problems raised concentrate on:

–   the counterpart situation, including the problem, that persons, who TAP are supposed to work with, are away on scholarships.
–   the fact that TAP for these and other reasons have to fill positions and do the job, where they had expected to have training or advisory functions only.

It is surprising how often TAP feel they are working in something like a transition period. 'Just now, the project has some specific and temporary problems, which makes it necessary to find ad hoc solutions'. Common reasons are that

— counterparts are away on scholarships
— counterparts have not yet been appointed or have just been transferred
— time gap between a leaving and coming TAP increases the workload for the remaining TAP
— all sorts of practical problems: break downs, procurement problems, negotiations, paper work etc. take time away from other more long-term tasks.

(ibid.: 113)

Recommendation:
System or schemes for transfer of knowledge interchange are very often lacking and should be developed.

(ibid.: 116)

Forss *et al.* (1988) found that local experts confirm the foreigners' performance. There was no evidence that a particular aspect should improve in order to facilitate the sharing of competence between locals and foreigners, and therefore there was no indication precisely which schemes should contribute to the transfer of technology. Local experts used arguments about equipment to explain that they could not perform in a similar way to the foreigners. Foreigners' allusions to scholarships similarly seemed to hide other difficulties. Either the interviewees could not express their other concerns or they were not able to formulate them explicitly:

The working atmosphere between expatriates and local staff is an important element in the transfer of knowledge. It is obvious that an atmosphere of hostility, envy and misunderstanding is not conducive to a transfer of skills, nor even to an effective job performance.

Very few project settings could be described in such negative words, but that does not mean, that there is a sense of common purpose, mutual respect and fine collaboration. The relationship between local and foreign staff is often friendly enough, but usually each party goes about his own business. If we conclude, that around 20% of the expatriates in our cases have a good and close professional working relationship with the local staff, we have not exaggerated. Seemingly trivial issues play a role in the organizational climate.

Expatriates have usually much better access to equipment. Vehicles are a particular problem ... Office equipment and office space is a similar but less frequently observed problem.

There are also differences in standard of equipment ...

It is also problems like these, which make the 'counterparts' weak, and gives the work counterpart a negative connotation in Kenya, Tanzania and Zambia.

(ibid.: 120)

Major problems of normative resentment appeared in the majority of projects. The petty problems of equipment could indicate deeper moral resentments which were not expressed:

In general the close contacts between aid agency and projects seem to be a precondition for an effective management of TAP. It is through a close integration of all aspects of the work that aid agency personnel have a chance to assess how well different experts are doing and what problems they have – and the same regarding local personnel. It is also through a close involvement with the project that they will have a chance to strengthen the performance of weak personnel, and if need be, terminate contracts and send home the expatriates. In general, the various monitoring reports that are used will not be of any help unless the quality of the agencies' integration with the project is good. Furthermore, if the quality of the integration is close and good, the reports are not needed anyway.

(ibid.: 123)

When an expatriate has completed a number of years on a project, he will return home. By that time, he will have gained considerable experience of the project, the country and the process of assistance. The danger is that this knowledge is lost to everyone but himself, and as he or she goes back to the ordinary activities at home even that memory will fade away. The problem for the aid agencies, and also for external organizations, is to document learning, to store the information, for newly recruited TAP and for its own personnel.

The Scandinavian aid agencies all have some sort of debriefing procedures whereby the departing expatriate should tell the programme officers of his experience. Usually they also write some form of report which is stored in the archives of the agencies. At times these reports are made available to new personnel – but most often they are simply forgotten. Why? Well, usually because they are not very good. The information they contain is trivial and if it should happen to be interesting in itself it is presented in a boring and haphazard manner.

There are also debriefing interviews, but they seem to be equally difficult to make interesting. These sessions may well be pleasant for the moment, both for the programme officers and the expatriates. But neither partner learns anything they did not know before – the value of the process lies more on the ceremonial side. It is not to be expected that an hour or so could yield any significant insights.

The sad fact seems to be that the potential for learning from the experience of individuals is seldom realized. Here the project cycle coincides with the individual's assignment, when projects come to an end they are forgotten and it is difficult to find any visible indicators for learning.

(ibid.: 126–128)

The patterns of implementation that we observe today are the results of ad hoc solutions to the problems of 'delivery'. But if the aid agencies are to be in control of development cooperation, they need to make rational choices between modes of assistance, and they need a conceptual framework to distinguish between such modes – as well as to discuss when a specific mode of assistance is applicable or not. It goes beyond the present report to develop such a framework, but our case studies have provided the grounds for identifying the need and we would propose that the organization of development assistance – over a broad range of projects – is investigated further.

(ibid.: 155), end of the Forss *et al.* (1988) evaluation study.

The last paragraph shows that development agencies do not understand what is going on in TA practice, which suggests that the Forss *et al.* evaluation was an exceptionally good one.[58] Clearly, Forss saw the closed character of the insider perspective and concluded that nothing the development agencies undertook to change would be able to open up the insider perspective. This is in agreement with our observation of the idiosyncrasy of each TA encounter, grounded in the contradictions between the local and the foreign perspective.

In addition to this conclusion, the proposals by Forss regarding TA planning are unlikely to result in changes to TA practice. The local experts' arguments about equipment did not seem to point to a reason for the failure of the professional collaboration. A better car, telephone or computer would hardly enable a technical expert to perform better. Nor would these facilities explain the resentments. But Forss did not take them seriously – he called these things trivial. Equally limited was their basis for suggesting modifications. Forss asserted that lessons learned cannot be recorded in agencies and stay only in the individual's know-how.[59] This analysis was well founded, but he did not suggest how to overcome this. We should add that, based on the case studies, one cannot improve the rudimentary learning within development agencies without changing the encounter between foreign and local experts itself.[60] Forss was well aware of the inherent limit of only studying evaluation. He had already shown that 'Planning and evaluation are two sides of the same coin, they belong together' (Forss 1985: 345). For this reason, Forss concluded convincingly on evaluation habits in Scandinavian development agencies,[61] but he had no means of suggesting possible changes.

Good local experts avoided working with foreign experts, with whom they did not share their skills, and therefore local experts could not prove their worth. Expert competence did not appear to be a factor for relative success. Forss's conclusion is congruent with the first latent process, the impossibility of sharing the instrumental core of technical knowledge.[62] A transfer of knowledge is very rare, and project implementation skills cannot be identified which would be congruent with exo-social exchange processes. All TA studied by Forss was in sub-Saharan Africa, where we found such an exo-social dynamic in Appui Technique. Regarding the interface, it is evident that only normative accusations appeared in Forss's work and no professionalism criticisms.

So far, we have seen that, in a particularly thorough evaluation of a larger number of TA projects, conclusions appear which reflect the latent processes.[63] More importantly though, Forss *et al.*'s (1988) insistence about evaluation problems and their cautious recommendations almost presuppose that social processes in TA practice remained hidden in the evidence that they had assembled.

Another possible method of looking for evidence of the latent processes is a comparison of evaluations from different development agencies. The differences between agencies can reflect their respective institutional 'sight defects'. Unfortunately, no other development agency has so far allowed an evaluation that is as deep as the study carried out by Forss *et al.* Possibly, other evaluations based on interviews with large numbers of experts from different TA projects[64] exist, but they remain confidential. The contradictions hampering TA practice are institutionally too problematic.

After the Forss *et al.* study, the next most thorough evaluation was produced around the same time (March 1988) by Wood *et al.* (1988) and concerns the USAID. How does this study differ from that by Forss *et al.* and what are the hypothetical evaluation conclusions? Wood *et al.* reviewed 212 evaluation reports: 'Overall, it appears that AID is not clear about exactly what information it wants in, or from, evaluation reports' (Wood *et al.* 1988: 17). Wood *et al.* also acknowledged the difficulties of evaluations, but their conclusions were less categorical than those of Forss. In the 'In-depth Analysis of Issues', two principal findings are important here:

> Technical assistance can make a signal contribution to successful implementation, institutional development, and sustainability by placing and supporting local staff in positions where they perform key functions.
> (ibid.: 40)

> Team composition was one of the issue areas most frequently commented on with respect to the success or failure of technical assistance ... For the sample reviewed, the performance, success or failure of a project, and its prospects for sustainability, were weakly related to any particular pattern in TA team composition ... Exceptional project successes or problems occurred with similar TA team sources and composition.
> (ibid.: 59)

Team composition appears to be very important, but there is no clear indication what the team composition factors responsible for the successes and problems are. The only clear positive factor is a central role for local experts. This does not appear in the hypothetical evaluation conclusions, but it is not in direct contradiction to them. Good-quality work by the local experts is quite possible according to the projections, but it is not noticeable when local experts work with foreign ones. In order to understand project practice, an evaluation should contain details of the actors' opinions. In the 212 evaluations reviewed, there was not enough detail for Wood *et al.* to reach a conclusion.

Reviewing USAID evaluations, Wood *et al.*'s analysis confirms Forss's research design. Because standard evaluations contain so little information about practice, it was necessary to interview participants to produce meaningful 'raw data'. But even Forss's interviews did not produce much conclusive evidence about TA practice other than the scarcity of lessons learned. It is quite possible that Wood *et al.*'s conclusion about the importance of TA team composition was just such a pseudo-conclusion, blindly reproducing TA practice. If team composition was the feature most frequently commented upon in USAID evaluations but the correlations between team composition and results were not consistent, then the conclusions from these 'evaluations' are deficient at best. Summing up the projections, the Forss study compares well with the defined latent processes. Studies that are not so comprehensive, such as that carried out by Wood *et al.*, contain less evidence for comparison.

These comments on evaluations would be incomplete if we did not look at IBRD evaluations. As expected, essentially the same comments apply to IBRD evaluations as to Wood *et al.* (1988). Therefore, there is no evidence concerning TA practice that would allow us to comment on the projection. But, reviewing IBRD evaluations helps us to understand why meaningful evaluations are so difficult to produce. Let us compare IBRD evaluations with the Forss study, and then let us speculate what the IBRD would say about Autogeneración. What appears in publicly available IBRD documents?

Our two case studies took place during a period of expansion. The IBRD's TA lending increased by over 60 per cent between 1990 and 1993, and then levelled out. That increase represents over \$US1 billion annually (OED 1996: 4).[65] These are substantial resources even for the IBRD, making it the biggest funder of TA in terms of dollars. At the same time, the IBRD revised its guidelines, resulting in new operational directives for TA (IBRD 1993). For these two changes to take place in parallel, there must have been considerable urgency. 'Outcomes have varied widely, but overall the efficacy and cost-effectiveness of TA has been disappointing, especially in sub-Saharan Africa' (OED 1996: 4). Given the urgency and the rather negative results, there are very good reasons for the Operations Evaluation Department (OED) to examine that period in detail in order to identify strengths and weaknesses:

New conventional wisdom within assistance agencies says that long-term advisors tend to undermine capacity building and thus that short-term consultants and twinning arrangements are a better choice. Especially in Africa, the so-called expert-cum-counterpart model has often failed to make lasting improvements in capacity. For their part, managers of poorly staffed institutions often prefer short-term operational assistance over long-term expert assignments for capacity-building; often counterparts are not available, or turn over too rapidly, or are insufficiently qualified, to benefit fully from the expertise available on long-term assignments.

Experience shows, however, that what matters most is the combination of the institutional environment and the qualities of the TA personnel. When the right experts are placed in settings conducive to training and skill transfer, as has happened in numerous countries, they have a significant impact. When these conditions are not met TA fails, no matter what the instrument used ...

Good management is equally vital on the provider's side. The World Bank's experience emphasizes that TA should be managed only by experienced staff. At the same time, the Bank's managers need to be sure that even experienced task managers get their attention and support as necessary. In the past, The Bank has too often appointed inexperienced task managers and tried to forestall problems by overpreparation and micromanagement, rather than appointing experienced, highly competent task managers and giving them freedom to make decisions.

(ibid.: 11–12)

The OED dilutes the conclusions drawn by Forss *et al.* (1988) regarding the operation of TA. While the OED is critical of foreign expert–local expert collaboration, its proposed lessons to be learned concern more the general environment of that collaboration. Where Forss suggested the introduction of systems for transfer of knowledge and more manpower planning, the OED claims that it is the combination of professional competence and institutional factors that matter most. The OED also stresses governance and ownership by local institutions. The key difference here is that the OED looks outside TA projects whereas Forss insisted on looking at what happens during TA implementation.

To assess the difference between OED evaluations and Forss's analysis, we use an OED evaluation that points to the key factors for success or failure. The OED has published a comparative evaluation between a failed TA event in Ghana and a successful one in Uganda (OED 1995b). In both cases, the elaboration of annual work plans was introduced in order to strengthen the flexibility of these TA events, supporting a process rather than a blueprint approach. While these annual work plans helped to negotiate the events during implementation in Uganda, they failed to do so in Ghana. The reason identified was that in Uganda 'Bank supervision was intensive, of high quality, and geared to a practical, results-oriented management approach' (OED 1995b: 4). The other major difference identified was that, in Uganda, a previous TA event had enabled potential weaknesses to be addressed, and the continuity and cohesiveness of the IBRD's country team and the local project management unit ensured that these lessons were effectively learned. In so doing, one element of TA practice was isolated as a key factor, but since that factor did not yield coherent results something outside TA must be responsible for success or failure. Looking outside TA then leads to contradictions. In this case, between the TA as a process which has to respond flexibly to local conditions and the need for closer supervision. A second contradiction is between the stress on strong management from the IBRD and the need for local ownership. The management vacillates between an 'arm's-length' style and a micromanagement style, combined with a need to continue with individual managers because the lessons learned cannot be identified. In sum, the OED evaluation does not get around the basic problem that Forss has described – how to overcome the idiosyncrasy of a TA encounter and isolate one aspect that would be independent of the individuals involved?

The idiosyncrasy of each TA encounter is also the cause of the important potential contribution by ethnology to evaluation methodology. A contribution which is unlikely to appear in the near future because evaluation science has yet to acknowledge the discussion outlined in Chapter 2, i.e. between anthropology of development and development anthropology. So far, only Rebien has referred to the appearing actor-orientation in evaluation work. Intrinsically, evaluation work will discover the agency of participants in TA events. Rebien applies Long's and Giddens's concepts (Rebien 1996: 145–150), repeating the debate we saw in development anthropology. Had Rebien not studied evaluations but the developer–developee encounter underlying them, possibly he would have arrived at the interface formats we have isolated (section 5.3).

The comparative OED evaluation functions more like an element of the blind reproduction of TA practice, something Forss avoided because, among other reasons, he did not want to be overambitious. The comparative OED evaluation, based on considerably less data, was too broadly defined and analytically overambitious (just as the TA projects themselves).[66] In publicly available IBRD evaluations, which refer to all TA, we find no general parameters capable of qualifying TA effectiveness. Possibly, internally available evaluations do distinguish between regions and industrial sectors, showing the specific differences between different types of interface or different exchange dynamics. Maybe there is evidence in internal evaluations that contradicts the hypothetical evaluation results of the projections, but we cannot verify this here. Closing this review of some particularly pertinent evaluation studies, the projection from the latent processes confirmed the plausibility of these processes. The projection cannot, however, be interpreted as a confirmation of the processes.

It is important finally to point to three conceptual problems for TA evaluations which are the result of technical knowledge. Financing a dam is quite easily defined, but training dam engineers is another matter. Let us briefly consider only the Mexican case study. Between 1948 and 1993, the

IBRD financed TA to Mexico equivalent to 6,000 man–years of experts (OED 1994).[67] Seventy per cent of these projects were evaluated as performing satisfactorily (ibid.: 2). We cannot judge whether the OED would consider Autogeneración as being satisfactory or not. The OED would not evaluate Autogeneración anyway because their analysis would take into account the whole $US450 million loan, of which Autogeneración was only a very small part – a practice with rather important consequences. OED evaluations comprise a cost–benefit analysis – does the outcome justify the resources?[68] Of the forty investments whose feasibility was analysed by Autogeneración, only four are being continued, but by experts who were not involved in Autogeneración. One of these investments provides benefits larger than the cost of Autogeneración because of the amount of fossil fuel saved per year.[69] But would these investments be realized without the feasibility studies? This is not certain. For example, a commercial bank, asked to provide a loan for such an investment, might favourably note that the IBRD funded such studies. Plant management might note critically that the Mexican government borrowed funds to support them. Therefore, the impact of the project is unclear because it depends on many assumptions for which there is no basis – the first conceptual problem.

What is a useful evaluation then? What can one get from an outside assessment that is meaningful in TA? At present, evaluations are operationally misunderstood. It is generally assumed that evaluations are too much of a management tool, failing to enhance the design and planning for TA. This is certainly correct for general development assistance, and more long-term impact analysis is necessary. This is a standard argument in evaluation science.[70] However, evaluating TA from outside does not even produce meaningful results for managerial purposes either!

What might be the managerial lesson from a cost–benefit analysis of Autogeneración? We have observed the managerial difficulties which resulted in Autogeneración remaining well below the project's potential as each expert contributed less than his/her experience (locals as well as foreigners) should have allowed. In particular, evaluations concluding a cost–benefit ratio cannot improve management. The very reasons for the effectiveness of a particular TA event cease to exist after the actors have departed – the second conceptual problem. Observing TA is possible only during TA implementation. If methodological tools are available to understand the event, it can be modified. No amount of planning, engineering or economic analysis and preparation can replace this. Having funded several thousand expert man–years of TA in Mexico alone, neither the lender nor the borrower seems to apply the managerial lessons learned. Current evaluation procedures regarding TA are as ineffective for general developmental purposes as they are for managerial purposes.[71] Conceivably, cost–benefit evaluations are still used for TA only because they were suitable for projects with physical components, of which TA was a necessary but small part. By itself, TA simply never has a viable basis for evaluations.

The severe shortcomings of cost–benefit analysis have been exposed many times. However, it is important to stress that there are specific reasons why TA is not at all accessible with cost–benefit tools. In Autogeneración, direct benefits would lead to an investment involving the experts who participated in it. In Appui Technique, the artisans' sales should have been in excess of the project's costs. As in most TA, the direct impact was only one aspect when these events were planned. The feasibility studies and the prototype machines were mere examples, whose worth was equally in their demonstration value. This is a habitual design approach, combining the direct impact with structural and dynamic impacts.[72] The latter is meant to enable local experts to learn from the event and apply the know-how to other contexts, designing a cogeneration system with different components (for example a gas engine instead of a turbine) in a different industrial process, or identifying another agricultural machine that is currently imported for Chadian artisans to manufacture. Local professional standards in energy engineering and in product innovation would benefit from TA. This dynamic impact ought to be evaluated because it is the core objective of TA. Our analysis of the two case studies highlights that the dynamic impact is even less clear than the direct impact – the third conceptual problem.

The experts used their experience, and the demonstration objects were indeed quite effective. In Autogeneración, foreign and local engineers went to a plant they had chosen, established operating parameters (ranging from weather patterns to differences between shift operation), assessed everything against benchmarks and then discussed many options, stopping short of industrial ecology (thermodynamic process integration over adjacent plants). The Mexican engineers could have adopted the foreigners' skills in identifying the key parameters for a feasibility study, had both sides managed to divide tasks and review them. Therefore, the dynamic impact was conditioned by the interfaces and, as a consequence, the project design and the dynamic impact were absolutely unrelated. The rare evaluation with some insight into TA practice supports the latent processes inferred from the case studies. This is not a verification, but unfortunately more is not possible here.

Despite the project design's vague orientation to a dynamic impact, the TA events were so complex that it is misleading to relate the dynamic impact back to the design. For an evaluation of TA, either the project design has to go into further detail about the tasks for actors (which is suggested repeatedly and always proves to be impossible) or the evaluation has to relate the evolving relations among the actors to the dynamic impact. The latter would leave the design out of the evaluation. This suggests that evaluations are often for the same reasons as vain as the TA projects. Currently, planning ignores the nature of the developer–developee relationship, and so does evaluation. Rather than summarizing what we learned about better evaluations, it is necessary to challenge TA evaluations as purposeful efforts as much as TA itself.

The review of the evaluations reveals the striking inadequacy of evaluation work on TA despite the volume of evaluation studies. Evaluation science has not yet discovered even major issues in TA. The development agencies are blind to TA practice and evaluations are also managerially useless. The idiosyncrasy of TA encounters has yet to be acknowledged and integrated into evaluation methodologies. Whenever evaluations in other fields (for example health) touch on the transfer of knowledge aspects, this serves quite different purposes in development agencies. Carlsson *et al.* (1994: 198) estimate that, at present, only around 10 per cent of cost–benefit studies in all aid provides substantial knowledge. He stresses that this share can be enlarged if these tools are improved and are used for participatory evaluations.

There are other lessons for evaluations. We include these (in the order of the projection), although evaluation procedures are beyond repair in many agencies. To avoid the misevaluation of the technical content of TA events, evaluators have to accept first of all that foreign know-how must remain different from local know-how while refraining from determining these differences. Evaluations cannot achieve or simply presuppose what foreign and local actors did not realize together. Evaluations in exo-social exchange contexts have to accept that the TA outcomes are inappropriate for the local context, as they were carried out partly in response to the foreign goals for the technical content. An evaluation meant to improve TA planning should focus on the degree of inappropriateness because potential lessons-to-be-learnt are there. Mobilizing indigenous knowledge or specific local know-how does not affect exo-social exchange processes. While this is always feasible, it does not affect developer–developee encounters marked by exo-social exchange dynamics.

Endo-social exchange processes indicate on the contrary that evaluations gain nothing from judging the appropriateness of TA outcomes; this is not at issue. In these contexts, actors made more progress than they thought, and evaluations must not take their frustrations into account. As the TA outcomes are deficient in ownership, they have a potential to be grasped. Instead of seeking institution-building measures,[73] the TA outcome can be valorized in imaginative ways. Such evaluations should preferably strengthen every positive activity that the actors identify.

The two structural formats of the interfaces are too uncertain to deduce a specific evaluation approach. Interfaces reflect the individual actors present. The interface process is more important for managing TA. Furthermore, evaluations cannot disregard normative accusations. These have to be dealt with even if they are not the root cause of the TA obstacles. Similarly, the competence judgements have to be taken seriously, even if they are overstated. As Forss *et al.*'s evaluation analysis showed, good evaluations do not consider these judgements on face value. Other evaluations, such as that carried out by Wood *et al.*, failed to detect that. While both normative and professional criticisms of and by TA actors are correct, they express something not contained within them.

There is no other suitable form of assessing the three latent processes before these are precisely defined. This being beyond the scope of an explorative study, we now shift the orientation. Until now, we have been clarifying TA reality, which we wanted to understand first of all. The case studies reproduced the internal coherence of both encounters. By introducing theory, we have already departed from this orientation because we had to reduce the complexity of actor behaviour. Nonetheless, this has proved useful in understanding TA events in general. Identifying latent processes enables practitioners to reflect on their work. Planners cannot use these results because they have no influence on these latent processes. The hypothetical evaluation conclusions (Table 5.1) help evaluators to re-examine already completed evaluations. But the above lessons for evaluations from the endo- and exo-social dynamics showed that these lessons are not far reaching. They can nonetheless avoid some hazardous evaluation conclusions.

The latent processes reflect the limits of the actors' efforts to use their know-how. These limits originate in their professional socialization and in the dynamics of their exchanges. We have not assessed whether these processes are specific to industry. Nor have we addressed the influence of tacit versus explicit or embodied versus disembodied knowledge because the processes are not intrinsic to specific technologies. Professional socialization and exchange dynamics are at work wherever technical knowledge and sociocultural differences between countries (and firms) mix. Commercial technology co-operation contains similar phenomena. In Chapter 6 we start to include contextual factors unrelated to technical knowledge and the particularities of development agencies are more prominent. In other words, we are looking for opportunities to enhance technical knowledge in the institutional context of development assistance.

We introduce theory in order to establish one unifying framework for TA which allows us to extrapolate from the cases. This is appropriate for exploring the options for changing TA practice. We thereby reduce the complexity of behaviour in a different way, primarily because actor behaviour is treated as more situated in a historical and social context. This is suitable for managers or planners looking for their degree of freedom to manoeuvre. We can include non-essential aspects as well as the latent processes. There are opportunities to improve project implementation, e.g. by assessing the coherence of secondary aspects. When actors receive a clear message from all the decisions they have made, their motivation and capability to perform improve. Evidently, the non-essential aspects can only be treated now, after the essential ones are established. Chapter 6 does not reflect on all of the constraints the latent processes impose on TA, but this should not reduce the pertinence of the management tools as long as there is a clear framework. We now move to TA *realpolitik*.

6 Technical assistance event management

TA management enables developers and developees to improve their work. This is a definition and an objective that we now describe as TA event management; events being the actors' efforts to agree on the content and purpose of their work. In the two cases studies, the content and purpose concerned manufacturing know-how and energy engineering know-how. We started with the proposition that technical assistance 'pyramids of sacrifice' can succumb or soften when latent processes become visible to their actors. If the latent processes explain what happened, they can also inform the actors. TA event management therefore consists of tools for the organization of projects that help actors to understand the latent processes. The challenge for TA event management is to identify such tools to the organization despite the actors' encounter being unpredictable.[1] The sociocultural complexity of technology is such that the encounter always overwhelms the actors. Thereby, it is unpredictable. The insights gained from the latent processes master this challenge. These processes are intrinsic parts of the encounter. Tools for the organization of projects that anticipate these processes help actors to weaken these processes.[2]

In examining management, the methodological focus of this study remains the actors' behaviour in the two TA case studies. Acknowledging that management is a relatively unexplored theme, we analyse the case studies and compare them with the most relevant document on management (IBRD 1993). As this document does not cover the whole field of TA, we refer first to a theoretical model of cultural distance. This model is applicable to all TA and allows us to avoid an uncertain extension of the management literature.

For evaluation science we looked at published work and at state-of-the-art work, while refraining from proposing alternatives. There is an abundance of documents on evaluation that have no relevance to TA, whereas the few documents on TA management have tremendous consequences. For management, we therefore propose an alternative in the form of a conceptual clarification of various *management tools*. We examine a sample of tools, determining their orientation and potential.

TA implementation is a hit-or-miss affair.[3] Planners, politicians, experts, etc. anticipate that they cannot amend TA in progress. An exercise in TA is initiated and afterwards, in the best cases, evaluations show whether objectives were met. Project implementation, i.e. what actually happens during the project, is a black box, the content of which cannot be known.[4] Neo-classical economists treat all technology similarly: as unknown entities with unspecified input–output functions. Production factors enter the black box (raw material, labour, etc.), products and waste come out, and all is substitutable and equivalent.

Why is TA management studied so little? Because, of course, it is so difficult to account for the unpredictability of the encounter; but there are other important reasons. Treating TA events as black boxes is not only convenient but it also reflects the dominant school of thought in economics: the dominant discipline in the aid business. Production factors in TA are expert man–hours (up to hundreds of man–years in a single project), their university degrees (how many people have qualifications such as Ph.D., M.Sc.), training for the local experts, support from the foreigners' home offices and, whenever possible, computer hardware and software. While the hit-or-miss method is problematic, the substitutability of the production factors is even more dysfunctional in TA practice. Without knowledge of TA events, and starting with a fixed budget, planners and experts enter inputs with no clear idea of the input's contribution to the specific expected outcome.

This is aggravated by the great concern for the little understood 'culture of dependency' fostered by TA (or the reproduction of that dependency). Planners sometimes even justify their practice of assembling inputs by arguing that cultural factors make it impossible to predict the input's contribution to the outcome. That which one cannot explain by examining inputs and outputs they label 'culture'. Neo-classical economists similarly label entrepreneurs' behaviour, beyond rational explanation, as 'animal spirits'.[5] Much of what is labelled as 'animal spirits' in economic data can be explained by market imperfection and risk assessment, factors that neo-classical economists systematically underestimate. Probably, postulating a culture of dependency fostered by TA serves to hide similar gaps in neo-classical theorizing.

Two other TA management traits confirm the impermeability of the black box: the existence of implementation 'trouble-shooters' and agency promotional practices. In each development agency there are 'trouble-shooters' who are parachuted into projects in crisis. These individuals have worked in their agencies for many years. They have mastered the agency's tacit rules – a mastery which is crucial for managing implementation. Second, there are elaborate developer hierarchies with promotional systems which have little relation to an individual developer's performance. Rewards or sanctions cannot be linked to something inside black boxes. At entry level, there are specialists, or officers, who provide specific subject competence. Next is a huge step to leader of the party. At that level, the hierarchy is based on project budgets. First, the developer is in charge of a $US50,000 or $US100,000 project, then a $US1,000,000 project and, finally, a project worth tens of millions of US dollars. It can take 20 years to climb up the hierarchy.

With operable performance criteria, developers could move down the hierarchy as well as up, but in TA practice this does not happen as career paths are predefined. Albert Hirschman's classic study has just been re-edited,[6] another indication that the black box has been virtually untouched for 30 years.

Opening it requires leaving the input and output variables out of the analysis and examining original internal variables. Black boxes appear to be black from one angle and quite different from another. Schumpeterian economists never believed in the black box called technology, and today's evolutionary economists have a feast putting colour in the box.[7] So far, we have used the case studies to identify exchange structures and dynamics, thus opening the black box of TA reality. Economists would now apply some sort of factor analysis. In examining management tools, we take a less positivistic route, starting from a theoretical framework then examining the fit of the empirical evidence. Again, we use anthropological theory as a framework to examine the role of management tools in the black box.

As we saw with evaluation science, the latent processes we identified do not appear directly in the literature. We later compare the *Handbook on Technical Assistance* (IBRD 1993) with the deduced management tools. In view of the scarcity of TA management studies, we do not classify the actual management in Appui Technique and in Autogeneración and we avoid other typology on an aggregate level. Comparing management modes between the case studies would also produce specifics of the two contexts. Instead, we pursue a more general route, providing an interpretative frame and then testing the implications for management. We thereby avoid literature reflecting the black box and avoid engaging in culturalist gap filling. Attempting to classify GRET's and Hagler, Bailly, Inc.'s management modes would produce a comparison of the insertion of TA into foreign policy (France and the USA respectively).[8] Although this would be enlightening, it is not pursued here.

Looking at the technological content of the case studies, we have already seen that the cultural specificity of that content is not a powerful explanatory tool. There is some specificity in the experts' heuristic habits regarding the technical content, and possibly some insights to be gained from these habits. However, more insight can be gained by looking at the sharing of the technical content, i.e. more technical content was shared than not, between the Mexican and US experts and between the French and Chadian experts.

We constructed the latent processes onto the case studies using the most pertinent theoretical approaches: interface analysis, global anthropology or the theory of communicative action. The latent processes cannot be inferred from the case studies alone because it is methodologically difficult to transform the interpretations of the events into objects of anthropological analysis. Each interpretation requires careful commentary on my fieldwork. Describing the latent processes, we have applied theory as well as generalized the interpretations from the case studies.[9] We use a combination of inductive and deductive reasoning to add variables to be tested in other contexts.

Deductive reasoning implies that one theoretical approach will bring the case studies together. One approach should be able to explain all evidence and, thus, increase the relevance of the conclusions for other developer–developee encounters. The benefit is threefold: management tools for defining the quality of the developer–developee encounter, revealing management choices hidden in planning and opening new routes to new tools. The first step, the definition of the dimensions of the encounter (section 6.1), is the most important and we apply Todorov's frame of analysis. To conceptualize tools, we have to reinsert the actors' subjective interests (section 6.2). We ask: 'What could have helped them in the pursuit of their interests?' Such an extrapolation moves decisively beyond the empirical basis of this study.

6.1 Dimensions of the developer–developee encounter

Instead of testing the wider implications of the latent processes, we move carefully through the implementation data to identify tools that enable actors to improve their encounter. Cultural distance being the source of their difficulties, we need a general but precise definition of cultural distance to extrapolate from the case studies to other TA events. Tzvetan Todorov has explored extreme examples of cultural distance by studying concentration camps, racist movements and emigration. Adapting Todorov's theory, we identify three independent dimensions in the developer–developee encounter.

For a seminal study on communication between cultures, Todorov chose the most radical human encounter with cultural distance:[10] the conquest of America. This encounter started with the unconscious, naive assimilationism of Columbus, in which Indians were seen as objects in the same way as animals and plants. The encounter evolved into an ideology of enslavement. Gradually, the colonists accepted that Indians could learn their language and become Christian. This earned the Indians an intermediate status, between object and thinking subject who can contemplate. Until Indian knowledge, values and religion are accepted as equal to those of the West, grasping the violence of Western assimilationism, many intermediate conceptions of cultural distance appear.

> To account for the differences that exist in actuality, we must distinguish among at least three axes, on which we can locate the problematics of alterity [cultural distance]. First of all, there is a value judgement (an axiological level): the Other[11] is good or bad, I love or do not love him, or, as was more likely to be said at that time, he is my equal or my inferior (for there is usually no question that I am good and that I esteem myself). Secondly, there is the action of *rapprochement* or distancing in relation to the Other (a praxeological level): I embrace the other's values, I identify myself with him, or else I identify the Other with myself. I impose my own image upon him; between submission to the Other and the Other's submission, there is also a third term, which is neutrality, or indifference.

Thirdly, I know or am ignorant of the Other's identity (this would be the epistemic level); of course, there is no absolute here, but an endless gradation between the lower or higher states of knowledge.

There exist, of course, relations and affinities between these three levels, but no rigorous *implication*; hence we cannot reduce them to one another, nor anticipate one starting from the other.

(Todorov 1984: 185)

TA encounters are marked by a contemporary form of otherness and we apply this definition of the three dimensions. Other aspects of Todorov's research are not used. Comparing individual actors' behaviour, it is possible to situate them on the value axis, the relational axis and the identity axis. Similarities between foreigners and locals highlight the overall dynamics of the encounter. An individual actor's choices reflect his/her conscious choices as well as his/her cultural limits. Discovering the Other, he/she discovers the self. Thinking about the Other, he/she forms social identity. Developers and developees are intimately related. Like the conquest of America, their encounter is radical, but in another way. They share centuries of history and feel overwhelmed not by their distance but by their proximity.

The value axis is the judgement of the Other according to the categories of good and bad, right and wrong, i.e. according to normative categories projected onto the Other. The relational axis of analysis examines whether the attitude towards the Other is one of separation or assimilation. The relational attitude is reflected in the behaviour towards both the foreigner and the local. When two people who have very different ways of working must collaborate, they have two options: they can combine their talents (assimilate the other) or work independently as much as possible (separate from the other). This relational axis is independent of the judgement of the Other, as either assimilation or separation can result from a good or bad value judgement. These axes are, then, orthogonal and do not necessarily influence each other.[12]

The identity axis,[13] which is more progressive, accounts for the recognition of the identity of the other, for an actor's desire to know the real other before him/her and understand his/her perspective. It is, of course, impossible to judge the recognition of the other according to formal criteria, but one can evaluate a will or a manifest effort to recognize the other as a subject. This axis hinges on comparisons between one person's efforts to recognize someone else's identity against another person's efforts and is, therefore, more progressive or gradual than the other two axes (Figure 6.1).

According to Todorov, all combinations of positions on these axes are possible but some are more likely, depending on the context of particular encounters. In fact, in the context of the case studies, it is theoretically possible to map these three axes into a single one. If an actor perceived and desired to understand the identity of the other, she/he was likely to endeavour to assimilate the other's methods and ascribe a positive value to the other. If an

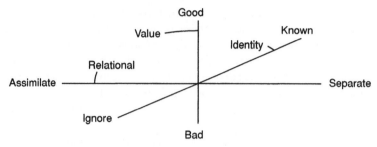

Figure 6.1 Cultural distance

actor manifested no desire to know the other, the opposite reactions tended to follow. However, although it is possible to map all axes into one, examining each separately expands the range of possible choices made by each actor and encourages a richer analysis.

Separating out the three axes will illustrate the logic of communication at work in the exchanges in Appui Technique and Autogeneración without recourse to strict formalizations such as those presented in section 5.4. We add to this theoretical framework certain elements of the sociopolitical reality of these projects, namely the relationship between France and Chad and that between the USA and Mexico.

Relocating the events in a global context clarifies the issues.

> Social action is not independent of a socially binding definition of the situation. For this reason, observable social action must be grasped from the perspective of the acting subject himself, a perspective that is removed from direct observation; that is, it must be 'understood'. The principle of subjective interpretation, or, better, of *verstehende* interpretation, concerns access to social facts, the gathering of data. Understanding symbols takes the place of the controlled observation, for the subjectively intended meaning is given only in symbolic contexts. Thus that principle defines the experiential basis of the sciences of action.
>
> (Habermas 1988: 54)

This implies that a particular aspect of these realities allows us to isolate these two relationships; relationships conditioned by a long history shared by these countries. For example, we propose that the very establishment of Appui Technique was marked by a mutism toward the other. However, we cannot demonstrate with certainty that this foundational silence was an expression of the colonial past and a symbolization of French (military) domination. Nor can we state definitively that the silence was an effect of the indifference to the local reality evident in the practical efforts of French assistance in Chad. In any case, these questions are all moot if we wish to compare this mutism with the similar initial conditions that manifested themselves in Mexico, where colonial domination goes back a century (before the Mexican revolution). Using Todorov's concepts, we maintain the primordial status of the exchanges between developers and developees.

6.1.1 Developee efforts of interpretation

Having defined the axes on which we will situate the encounter, we now refer back to the two case studies. The task for TA event management is to assist developees, Chadian artisans or Mexican engineers in improving their comprehension of the encounter. To express this using the actors' terms, we summarize their interpretations, already presented in Chapter 4. Along the three dimensions of cultural distance, their interpretations were close, i.e. all foreigners and all local experts interpreted the distance to the other group in a similar manner. Once the developee and the developer positions are identified, we will again analyse the effects of the observer role (see section 3.2) to describe how they attempted to change the cultural distance (in section 6.2).

In Appui Technique, there was a sense among the Chadians as well as the French experts that France had broken its agreements with West Africa by failing to take responsibility for its assistance efforts. In Autogeneración, foreigners and Mexicans accused the other and their relations were antagonistic. And yet, the enormous economic, social and political influence of France on Chad is far stronger than the influence exerted by the USA on Mexico.[14] The web of polemics may explain in part this 'reflexive inertia' from which TA actors suffer. We now use Todorov's concepts to compare this web among the developees in Chad and in Mexico.

Appui Technique

It is important to keep in mind Chad's economic dependence on other countries. Most of the state's subsidies come from import taxes and outside assistance. These data must be read in terms of Chad's colonial past and present political situation. Colonial interest in this region of Africa was historically very limited. The sixteenth-century Islamization of Chad was limited to the northern part of the country. The conquests of the proslavers were largely determined in battles with French colonizers in the early part of the twentieth century. The forced introduction of the cotton crop, instituted under the protection of the police, broke down the collective work habits in the south, where private ownership of land had never existed. The Chadians' economic sustenance was based on millet, sorghum and the peanut. The northern populations were, for the most part, untouched by the colonists' influence elsewhere, and their way of life was changed less markedly.

Since Chad's independence, development agencies have launched economic development programmes, aimed mainly at cotton cultivation.[15] In Chad, industry exists only as an agricultural phenomenon. Suffering continually from poverty since its independence, the country has been the recipient of projects from the ILO, the UN Food and agriculture organization (FAO), the *Coopération Française*, NGOs and many others. The national textile company, another colonial legacy, declared bankruptcy in the early 1990s. The old model of economic development imploded under the weight of outside changes, such as the cotton price on the world market, and exploded on the inside as a result of the incapacity of the post-colonial government to produce a national policy. Today, lessors find only empty offices and incompetent managers where there was once thriving industry.

The administration of the Chadian government charged with the supervision of Appui Technique is a good example of this qualitative decline. It has no budget, not even to buy paper and pencils. Civil servants' salaries, often paid months behind schedule, no longer feed their families, and government employees are often forced to work in other jobs on the side to make ends meet.

According to my interlocutors, the vital aid furnished by the foreigner is an admissible, even desirable, enterprise for Chad. While the French military has at times determined who holds power in this country, the military influence is unstable.[16] Over the last 15 years, the capital has endured a constant threat of combat, latent but ever present. The most recent president established a totalitarian regime, which was tolerated by France. This regime never put France's power in jeopardy, but it committed atrocities and maintained its own power through fear.

These remarks help us to understand the context in which the West African social Other is constituted. In spoken Arabic, the Other is the 'nasarra', but the artisans all agreed that this term was no longer adequate and that it was necessary to see the White man in another light. Yet, at the same time, they failed to reconcile the Other with the real foreigners that participated in aid and assistance. They accused the foreigners of causing the failure of the so-called development initiatives. In this sense, the Other was negative and his 'projects' were a waste of time. On the other hand, the artisans were conditioned to see the Other also as a big brother or helpful partner. Read this way, the White influence was welcome, desirable and even constituted the only hope for their future. *'We want the French to exploit us, and even that they exploit us more so that we can move forward'.*[17] This double perspective on the foreigner was tremendously useful, as it enabled the experts to fail in reality but survive symbolically. Despite the failure of aid and assistance, the White man (in this case, France) remained the hope for Chad's future.

Instead of assessing the complexity of the nasarra symbolism in general, we now look at its articulation in the exchanges studied. We have already identified salient differences in the behaviour of four artisans in Figure 4.1 (p. 64). Having described the latent processes that are independent of the individuals, it is possible to go beyond behaviour and assess the complexity of the meaning of the encounter.

The artisan who best knew the Other was Osama, who had travelled in Europe. However, his knowledge did not earn him a privileged status among the other artisans. For Osama, the Other was a wise and powerful judge worthy of emulation. Consequently, in his mind, the authority to judge belonged to the Other alone; a fact which formed his perspective of the know-how of the experts. Technical advice had an absolute value and was, therefore, unquestionable. In this sense, the superiority of the French was characterized by severity. The process of undoing this superiority had to begin with an interrogation or outright denial of the legitimacy of the nasarra's omnipotence. As these options seemed contradictory and the choice between them too difficult, the artisans simply avoided all discussion about Chad. In general, they could not express their political opinions. A few sometimes dared to voice a call to action – *'We have to stop this nonsense in Chad!'*[18] – and decide to follow the advice of the Other, or refuse it entirely.

Thus, for the artisans, the value of the Other (Todorov's first axis of analysis) and the relational attitude (the second axis) did not function independently. Since they reacted to the omnipotence of the foreigner without qualifying this authority in any way, their relational attitude was predetermined. Rahman and Mohammad clearly hoped to assimilate the experts' knowledge and encouraged the others to do the same.[19] They did not condemn the presence of the foreigners, affirming that the 'projects' were well executed and that the White men were only doing their job. There was doubtless a discrepancy between the Other that had once appeared as the colonist and the contemporary Other present in the foreigners. But, despite the differences, the artisans compared the two and were therefore disinterested in the actual other. Rather than dealing with the actual foreigners, the artisans reacted to the social Other. In some sense, they interacted with a virtual foreigner. The actual presence of the foreigners condensed the Other and fixed the imaginary other without recourse to the particular foreignness before them. The artisans maintained their social Other despite the contradictory behaviour of the actual foreigners. In the simplicity of these two positions (the only two possible), we witness the symbolic force of the nasarra. Differences in the relational attitude could not appear.

To conclude the Chadian case, we now have to assess the identity axis, the recognition of the other. For Ngerbo and Osama, the authoritative omnipotence of the foreigner was aggressive; this was an aggression that caused the Chadians to insulate themselves from the judgement of the Other. Osama was more explicit in his rejection than Ngerbo, who demanded the foreigners' assistance but made no effort to profit from the time he spent with them. Osama refused to open a dialogue with the experts and vehemently defended his own methods. Even more than Osama, Ngerbo acted out an exclusively exo-social process, in which he identified the foreigner as the singular source of power without questioning this power's character. He ascribed almost pure legitimacy to the foreigner (similar to the life-force of the *Sapeurs* that Friedman described; Friedman 1994: 105–108). Born in the southern part of Chad, Ngerbo was rooted in a Christian milieu in which the equality of all men and their brotherhood in community were fundamental principles. Osama was born in the most easterly part of Chad (Abéché), a region with a very rich Islamic history and a society structured by a caste system. They both moved to the capital, N'Djaména, during their adolescence and lived with their families in different neighbourhoods, which were organized and separated according to ethnic origin. They did not know each other before their engagement with Appui Technique and, despite their work together during the project, they did not develop a personal relationship. The only thing they had in common was their agreement (often made between the people from the south and the people from the north of Chad) to keep their distance.[20] Their agreement applied only to the foreigners' presence. Because of this limit, they were unable to ascribe other meaning to their work together. For 2 months, they manufactured a grain mill, passively following the experts' explanations, but never spoke between them about it (or the purpose of doing so).

Regarding the recognition of the identity of a foreigner, we can draw parallels between the behaviour of Mohammad and Rahman, on the one hand, and Osama and Ngerbo, on the other. Mohammad and Rahman easily constructed the identity of the White man. Recognizing their individualism, which prevented them from forming mutually deep bonds, seemed to enable their construction of the Other. This construction was not based on elements borrowed from the foreigners' discourse, but on their distant gaze directed towards the White men they met in the workshop. From this subjective perspective, it derived its coherence. Rahman and Mohammad could not recognize, as Ngerbo had, that the majority of the experts working for the IBRD were Anglo-Saxon. They could only see foreigners through their own realities: the distinction between the English and the French was irrelevant. Ngerbo and Osama succeeded better in recognizing the foreigners as individual subjects.[21] Occasionally, Osama borrowed elements of the foreigners' discourse in order to demonstrate his equality with the experts. Osama and Ngerbo spoke about their families at work and introduced me to them. Rahman and Mohammad kept their personal and professional lives separate. Osama and Ngerbo could not understand this attitude and cruelly poked fun at Rahman, in particular, considering him to be a '*damn fool*'.[22]

The recognition of the foreigners' identities, in the way that Ngerbo and Osama were able to, could not become an object of discourse. Their reluctance to articulate an appreciation of the individual foreigner is evident in the artisans' response to Chauvel's (a foreigner living in N'Djaména) visit to the yard when many artisans and experts were present. The artisans said they knew that Chauvel had become 'Islamized', having seen him at the mosque. However, in front of the other foreigners, who condensed the Other, the artisans refused to speak to Chauvel. Even when he addressed them in Arabic, their response was timid and evasive. The contradiction between his behaviour and his skin colour blocked their recognition of his proximity. No one knew what to make of him. This indicates that the developer's image was determined by the nasarra role. The White man's know-how (the reason for the artisans' presence in the yard) disappeared when the status of the foreigners became contradictory.

To sum up, the value of the Other was the defining element in the exchanges with the foreigner in Appui Technique. We observed that the positioning of the developee's identity determined the exchange dynamics (pp. 85–87), and here we see where this positioning originates. The historical importance of the Europeans in West Africa became part of the symbol of the encounter between developers and developees. Much more ought to be said here; in fact, the whole complexity of Todorov's account *The Conquest of America* will one day be elaborated regarding West Africa. But this is beyond our scope here.

Autogeneración

What does a similar reorientation of the exchanges in Autogeneración yield in the larger context? For the Mexican expert, the Other was the gringo – an epithet applied to all those who hailed from the USA. According to Octavio Paz, this term also has a figurative meaning: 'the United States is the image of everything that we are not'.[23] The gringo is then also the giant from myths and fairy tales, who can be as gentle and kind as he is violent and destructive because he does not know his own strength. One can fool him, but the risk is great as his wrath is mighty. The fascination with the gringo is equally ambivalent. He is the enemy of Mexican identity, but he is also the model of all that Mexicans want to be. In fact, this Other is inseparable from Mexico.

Besides the figure of the gringo, another menacing image informs Mexico's conception of otherness – Spanish colonization. An eternal doubt hangs over Mexican identity – are the sons of Malinche Indians or Europeans? The fear born of the fundamental instability of Mexican identity is directed towards the gringo as the embodiment of all that is foreign and occidental. The Mexican must always reject it or agree to use this figure as a measure of his own worth and as a foil for the stabilization of his own identity. When he fails in this task, he succumbs to the threat of recolonization – of the 'reconquista'.

Since the Mexican revolution of the 1910s, and to a greater extent since the nationalization of industry by President Lázaro Cárdenas in the 1930s, Mexico has been free to follow its own course, buffered from external influences. Because of its size (a current population of nearly 100 million) and the revenue from the exportation of oil, Mexico's social and political transformations have been largely determined by endogenous processes.[24] The 'Party of the Institutionalized Revolution' (PRI) has controlled the government since the revolution, gaining the support of farmers, workers and the middle class. Only the students and the intellectuals, and recently the Zapatista movement among the Indians of Chiapas (the southernmost region of Mexico), take the party to task. The US influence is limited to the migration of Mexican workers to California and Texas. Technical and financial assistance brought in from the outside has no influence on the Mexican economy.[25] The political process within the Party is a reflection of larger sociological processes in Mexico, even 50 years after its creation and despite its undeniable oligarchic elements. The Mexican experts of Autogeneración cared more about the national sovereignty than whether sociopolitical problems were actually addressed. Today, a state of unity in Mexico is synonymous with an endogenous existence, even as the tensions between an Indian and a Spanish identity strain its possibility.

The foreign engineers who participated in Autogeneración could easily become the site of the superposition of the old colonist onto the gringo. They came from US companies and brought with them technical knowledge that was to become the new measure of success. In short, they represented the power of the gringo, the power of the First World. Rather than look upon the gringo as an imperialist, Mexicans regard him as an ancestor. The construction of their identity is first a function of Spanish influence and later a function of the legacy of the gringo.

The superposition of the conquistador onto the US expert is captured in the term 'malinchismo'. The term bears witness to symbolic work in progress, work which does not find sufficient grounding in Mexico's past and which, therefore, instigates the construction of an Other. The term is a misrepresentation of the superposition, however, because it posits an historical explanation for it (Mexicans betray Mexico as Malinche did) rather than exposing the absurdity of historical explanations for such phenomena. The dynamics of the construction of the Other are such that the superposition of the threat of colonization onto the gringo becomes automatic.

In the current political climate, the word malinchismo directs the dynamics of construction towards the modernizer. Octavio Paz illuminated the contemporary function of this term in his marvellous anthropology of modern Mexico:

> Cuauthemoc and Doña Maria are thus at the same time contrary and complementary symbols. And if there is nothing extraordinary about the cult that we profess for the young emperor – the only hero worthy of art, the image of the sacrificed son – then we must not be surprised by the curse that hangs over Malinche. To this curse we owe the partisans of Mexico's opening up to the outside: the real sons of Malinche, who the *Chingada* incarnate. And so the closed world, opposed to the open world, reappears.
>
> (Paz 1961: 86)

The Mexican experts that spoke directly of malinchismo during their interviews (Juan and Vincente) invoked it to explain to me (a foreigner) the contradiction apparent in my ethnologist's role. From their perspective, I sought in my research to explain why it was difficult for the foreigners to help the Mexicans to learn what they knew. Their malinchismo allowed them to resolve the apparent contradiction (coming to Mexico to do something without the means to accomplish it), each in his own way. The term merely exoticizes a Mexican custom. Its usefulness as an accusatory charge in journalism and as a clarifying metaphor for Juan and Vincente lies in the inverted correspondence between what one hopes to explain by it and the superposition to which it ultimately refers.

The force of the construction was such that the reaction to the foreigner was singular, total and unaffected by the temporal evolution of the relationships. The foreigners spoke Spanish very well and were at home in Mexico, a place they knew well. Then they were also extremely ambitious and contemptuous of the joint work. The diversity of their ethnic backgrounds lent itself to their designation as gringos. María's reading of this book (section 3.3) corroborated these observations. She saw in it a confirmation of the domination exercised by developed countries over Mexico. She also saw the inability of Hector (the representative of the Mexican government) to demand real work from the foreigners and to give the Mexicans a chance to prove themselves. Her relationship with the foreign experts remained distant and rigid. John and I agreed that, of all the Mexican engineers, María learned the most about the technical content of the knowledge and the energy sector in Mexico. Although the foreigners as well as certain Mexicans (María included) actively sought and occasionally managed to combine their efforts, the perception and value of the Other remained for the most part culturally determined at Autogeneración.

Contrary to the conditions in Chad, the value axis and the relational axis were largely independent in Mexico. María, José and Ramón shared a vision of the Other, the developed gringo – strong and intelligent but cruel and domineering. The Other remained frozen on the value axis. On the relational axis, the positions of the local actors were diverse. Throughout the project, Ramón thwarted the foreign engineers' efforts by refusing their methods. María and José, on the other hand, were ready to follow all of the foreigners' recommendations. For both Miguel and Severino, the Other was yet more negative – a cold, barely human, presence. Yet their relational attitudes were quite different. Severino put all of his energies into improving the results of the others, while Miguel tried always to work independently. The actors' positions can be located along the axis between full involvement and independence, or assimilation and separation (see Figure 6.2).

There are many factors that account for these differences, including technical ability, professional experience with foreign companies and the hierarchical position of the Mexican expert. Figure 4.1 for the Chadian experts (p. 64) is similarly illustrative, but fails to reflect actor behaviour in its complexity. The graphical presentation cannot add to the argument we pursue.

On the axis of the identity of the real other, the Mexican experts' perception of the US engineers was very limited. María had no frame of reference for the remarks about Mexico that Bill made or for Joe's jokes,

Assimilation ◄──────────────────────────────────► Separation

Geraldo José María Aníbal Silvio Severino Lorenzo Juan Carlos Miguel Ramón

Figure 6.2 Relational attitudes of the Mexican experts

which seemed to her to mock Joe's Peruvian origin (which they did not). Ramón and Lorenzo saw the foreigners as 'fregónes' and 'real men', and ignored their individual differences. They responded little to the personalities of the US engineers. These foreigners' professional affinities seemed to justify this attitude. Despite the fact that the US engineers of Latin American origin (Juan and Joe) and those who had never worked outside the USA or France (Bill and Jack respectively) displayed very different attitudes towards Mexico and each other, the Mexican experts remained indifferent to their diversity.

In Mexico, then, the value axis did not structure the exchanges. Rather, it was their relational attitude that informed the Mexican engineers' exchanges with the foreigners. As for the identity axis, there was only one single Mexican position – their conduct was absolutely uniform. The value of the Other cannot explain this unity since, unlike the omniscient judge, the gringo is negatively marked. As a result of the gringo's aggressiveness, the Mexican experts questioned his power. That is to say, the superposition of the conquistador onto the gringo does not explain the singular position on the identity axis because only some of the Mexican actors invoked malinchismo or other national references.

6.1.2   Developer efforts of interpretation

Having schematically described the local actors, we now present the foreigner's side. First, we examine the French in relation to what was said about the nasarra, and then the same for the US engineers. We still remain in Todorov's frame of analysis and fill it with the context and events of the case studies. Completing this for all actors, we can then identify the interdependencies of the foreign and local perspectives. The interdependencies of the foreign and the local interpretations of cultural distance are the key to changing the encounter (section 6.2). This can only be done by the actors, and we look in detail at their efforts. To compare an encounter, it is not feasible to isolate developees from developers and relate the developees from one case to those of another.[26] But the interdependencies between developers and developees can be compared between cases.

Appui Technique

For the foreigners at Appui Technique, the Other was a beneficiary of assistance and a former colonial subject. The figure of the beneficiary grew out of the context of the political and economic crisis in Chad. One saw the traces of war on the walls in the heart of the city. One felt the tensions around the city in the shanty towns. Aid and assistance were noble gestures that the Chadians accepted with gratitude. The local experts recognized the artisans' plight as a real problem of survival. This plight justified the experts' offer of aid and the call for assistance from the outside. But assistance also compensated for a handicap that they inherited from ancestors held down by colonial domination.

The harsh conditions of the artisans' lives and Chad's historical legacy of dependence made up the French reception–construction of the Chadians. The discrepancy between the constructed Other and the real local actor was perhaps as poignant for the French as the gap between the local vision of the White Other and the personalities of the French present in the workshop was for the artisans. *'The rural Africa which remains friendly and which you try to discover ...'* was not a part of their daily experience.[27] Even Martin's and Jacques's 15 years of experience in Africa did not teach them to receive the individuality of the local actor. On the contrary, their professional experience probably dulled their sense of novelty, reinforcing the divide between their vision and the reality before them. In order to perform as an expert in assistance, a foreigner must impose this distancing. In his discussion of rural development, Jean-Pierre Chauveau explains how development credos discourage acknowledging the local actors:

> ... much more than on the codification of professional criteria specific to the practical action of development itself, the culture of development has based its professionalism on the codification of the effects, invariably perceived as perverse, which oppose this action.
>
> (Chauveau 1992: 27).

For the foreigners, separated as they were from the other that they were addressing, the identity of the developee became an effect of the past. The Other was the *roublard* they had often encountered and who symbolized in many ways the pitfalls so inevitable in the development experience. As such, the Other condensed numerous negative characteristics against a backdrop, positive but blank because this legacy of dependence could not be abstracted. In fact, the residual manifestation of colonization was not stigmatized. The colonial high-chair was off-set by the useful knowledge brought by the colonial administration, which was still recognized for its utility. Among themselves, the foreigners made the connection between colonial knowledge and the knowledge of the technological developer. Only the methods were meant to have been revised.

The foreigners could not make sense of their presence in Chad; despite all the aid, things had become worse. They had little contact with other foreigners and, during the encounters they did have, they spoke little of their work. An interrogation even into appropriate technology set off a full investigation of the French presence, especially the *Ministère de la Coopération*.[28] The tremendous difficulty of putting new, more enlightened methods into practice has in part caused the crisis of the developer who aspires to be a reality rather than simply 'doing projects' and failing. Consequently, the Other remains fixed in the deceptions of the developer.

Not surprisingly, the first artisans whose names the foreigners learned were those who responded best to their technical knowledge. Even after several weeks, the foreigners only knew a few names by heart. They dared not ask the artisans about their ethnic origins, a perfectly non-technical subject; instead, they directed these questions to me. All that concerned the social conditions, the political situation or the ethnic conflicts in Chad remained part of the archaic mystery of the Other. Among the foreigners, stories of earlier TA experiences seemed to focus always on the distinction between the actions of the foreigners and those of their interlocutors. The interlocutors could resist the foreigners' advances, or they could transcend their local context and embrace the foreign knowledge.

In summary, there was a strong similarity between the foreign Other of the Chadian artisans (Osama and Rahman) and the local Other of the French (Martin and Jacques). They were both fundamentally positive images that, nonetheless, failed to bridge the gap between stereotype and reality. In the space of the encounter, the condemnation of the Other by the foreigners was strong. They expressed it as a recognition of their (the artisans' and the foreigners') divergent views on the moral standards for the project. The foreigners considered the Chadians to be dishonest. Furthermore, they disagreed on the appropriate ethical conduct in their work with the artisans. Pascal was glad to leave the project because he resented Jacques's heavy moralizing. The latter was no more pleased with Pascal's Christian ethic.

The foreigners manifested only a limited capacity to recognize the identity of the artisans because they were unable to discern the Chadians' interpretations. Also, the symbolic force of the image of Chadians as 'uncivilized' was more destructive to the foreigners' perception of the Chadians than the latter's supposed moral poverty. In other words, the mystery of the Other was stronger than the real difference of the other. The foreigners were trapped at a distance from the artisans. This placed the foreigners closer to the resistance end of the relational axis. Only Pascal among the foreigners and Rahman and Mohammad among the artisans sought to integrate the methods of both groups. If these artisans could envision collaboration it was because theirs was a different Other from that of their colleagues. Their flexibility on the relational axis, an opening *vis-à-vis* their experience with the foreigners, was the result of their valorization of the Other. Similarly, Pascal stood out among the foreigners for his desire to collaborate. He was able to move further along the identity axis and open himself up to the personalities of the Chadian artisans.

Autogeneración

For the foreign engineers in Autogeneración, the Other was a composite of difference due to the diverse origin of these engineers themselves. John, Bill and Jack were very intrigued by rural Mexico and the naturalist myth of the simple man of the indigenous culture embodied in the Indians and their ruins. John was wildly enthusiastic about Guatemala (as was Jacques about Madagascar and Martin about Burkina Faso). Jim, Joe and Ben did not share this utopian vision of the 'underdeveloped' but professed rather a resentful wariness of the obstinate, prideful, nasty native. Neither of these exaggerations represented the reality of the Mexican engineers.

In their interviews, the US experts of Latin American origin incorporated social factors in their understanding of the Mexicans. For example, Joe, who was awarded his degrees in Peru, held that the Mexicans could only grasp the 'foreign technology' with great difficulty, as they were not aware of the cultural value attached to it. These experts tended to interpret the weakness of the Mexicans as a function of their professional habits, whereas John and Jack (who studied in the USA and France respectively) attributed it to various individual incapacities. John, in particular, went so far as to charge Hector with 'incoherence', despite his respect for the latter's earlier success in launching a consulting firm before joining the Mexican Energy Agency. Jim, from Argentina, managed better to take the Mexicans' skills into account, but nonetheless felt that the foreigners had to *'wield a stick behind the Mexicans'*.

The foreign engineers' relational attitude was clear: there was no combination of Mexican and foreign experience. The Mexicans had to understand their limits and, in principle, the foreigners felt the Mexicans were capable of doing so. All the same, the foreigners were frustrated for not being able to get them to do so. This relational attitude was shared among the foreigners. For this reason, the foreigners in Mexico asked me fewer questions during their interviews than the French.

Throughout Autogeneración, the foreigners knew that their ability to understand the motivation of their Mexican colleagues was the key to their success. However, this knowledge did not make them consider the personalities of their interlocutors; hence their insistence on giving Severino the same attention they gave to José, although the former was incapable of understanding even the nature of the technical analyses that José produced. The positions of all the foreigners were the same on both the relational and identity axes. They could not experience the Mexicans as individuals and this never changed. Their position in the project determined their way of relating to the local actors more than their origins, their pasts and their communication skills. In short, the value of the Other did not factor into their relationships with the Mexicans.

We have seen that the respective interpretations of developers and developees were very close. Especially on the identity axis, the four groups had almost uniform positions. On the relational axis, only the Mexican engineers had diverse attitudes. The significant relational differences among the Chadian artisans remained hidden even among themselves. The value axis showed minor variations within the four groups. Introducing the historical context of their encounter reveals that their interpretations of cultural distance were not related to their individual life histories. This is part of a foundation for extrapolating these interpretations to other TA encounters and a basis only for classifying management tools.

At this point, we cannot consider possibilities that, for example, would weaken the link between the relational and the identity axes for the foreigners in Mexico. It would be misleading to use the historical context as a sufficient condition for the actors' interpretations. Comparing the historical context of Appui Technique or Autogeneración with that of another TA encounter allows us to classify tools but it cannot be used to identify fundamental management goals.

Instead, we keep the three dimensions, their pertinence well established, and see in the next section how these interpretations are reinforced mutually. Then, we return to individual actors to identify their attempts to change the encounter. Considering these attempts together with the interdependencies of the actors' capacities, we have a viable empirical basis to identify management goals.

6.1.3 Intersubjectivity and the management landscape

While much of the aforementioned was quite independent of the actors, there were specific aspects with crucial interdependencies between the local and the foreign perspectives. In fact, these interdependencies made the actors' efforts so frustrating, always leaving an impression that the cause was partially their own shortcoming.

The transformation of the foreigners into protagonists was a result of their struggle against the Other (section 3.1). For the foreigners in Mexico, the struggle was fortified by a competition among the US experts, whose professionalism was reinforced by the limits of the Other. The French in Appui Technique lived a process informed mostly by France's past in Africa. John explained in the Washington office of Hagler, Bailly, Inc. that in Mexico it was difficult to obtain services so simple as having a document typed. Simple needs became complicated obstacles in this world. On the other hand, Martin of GRET told anecdotes in Paris about the functioning of the Chadian workshops which implied that much more could have been achieved by the artisans, for whom a well-soldered machine was meaningful.

The situation in Chad was marked by the colonial past and its precarious present relationship to France. French domination would have made itself felt in Appui Technique even in the absence of Chadian actors. The NGOs, the IBRD and the French aid agencies seemed to be omnipotent forces. The local actors were blocked in their deconstruction of the nasarra (necessary to their advancement), as Osama and Ngerbo tried to do, because such deconstruction would have required an interrogation of this figure's presumed role as a civilizing force. Still, the foreigners did not distinguish the symbolic weight of France from actual foreign aid and technology available. This deconstruction began long ago and is advancing because of the severity of the current crisis in Chad, although it has only now become truly engaged.

Until recently about half of the Chadian artisans were indifferent to the possibility of choosing or changing their relational attitude (position on the relational axis). In the context of Appui Technique, the foreigners again professed an alleged role of French influence. And yet, the need to believe in their myth placed them in a difficult position – the lived situation on the project showed them the illusory nature of this coherence. In response, they sought ways of distinguishing their action from the ones typically associated with a French developer, the *coopérant*.

The extent to which we can study the foreigner's role is limited in the absence of feedback, available only if the artisans assert themselves in their relationships with the foreigners. The deconstruction of the nasarra is the *sine qua non* of the artisan's reaction to the foreigners. Such a reaction would further allow the foreigners to act in turn towards an individual Chadian. The blockage, we have noted, is an interdependence between the coherence of the foreigner and his image of the local actor. The foreigners remained trapped in the crisis of their identity by boxing the local actors into an equally restrictive category. Such was the debilitating reciprocity of cultural distance (alterity) in the lived experience in Appui Technique.

This reciprocity was stable to the extent that the foreigners depended again upon their coherence to deliver them from the bind they were in, and in which they were called to act. However, during their private lunches among themselves, the foreigners asserted that the corruption in France had been as important as the corruption in Chad, or that the current development efforts were absurd because they continued to put into practice old unproductive ideas. Objecting privately to the purported coherence in the action of development, the foreigners tolerated the obvious contradiction between their theoretical 'aid' to Chad and its ultimate inefficiency in their interaction with the local actors. In fact, they bore it better than they bore the gravity of the country's socioeconomic crisis. Whenever they were with the artisans, the foreigners fell back on the image of their omnipotence and silenced their fears.[29]

The foreigners thus reinforced their cultural distance (alterity). On the other side of the interface, the Chadians invoked the nasarra in their interaction with the foreigners, either to catalyse its deconstruction or simply to react to their behaviour. The latter response, typical of Mondai, for example, involved submitting to the foreigner's power, and accepting the ideal of the White man as the driving force behind Chad's advancement.[30] In the presence of the nasarra, the goal was to remain as passive as possible so as not to hinder, in any way, his goal of development. The deconstruction of this negative dialectic (for example in the way that Osama was able to do) was begun with the coherence of the foreigners being contested. Osama's first gesture as a participant in the project was to approach Jacques with a management mistake that Jacques had made. But Osama's success in noticing and denouncing the mistake (which pleased the other foreigners) ignited in him the expectation of an equal relationship with the foreigners – one they could not grant him in the end. As a result, Osama was both appreciated and scorned by the foreigners. Osama's frustration led him to confront head-on the foreign domination, a reaction that later caused him to slip back into a passivity similar to that of Mondai and others. Neither the foreigners nor Osama were in a position to profit from constructive conflict, particularly as there were other artisans present who did not participate in the process. The cultural distance (alterity) of the local actors was stabilized in this way. We cannot comment on the pertinence of this stabilization in other types of projects which evolve in another institutional environment. However, the available results of other studies do show that the logic of communication at Appui Technique was of a novel type.

> What remains is to explain the permanence of the culture of development's own 'cognitive map', and notably the way in which the critical function, inherent in a rational bureaucracy, not only spares but also reinforces the idea-values of development culture.
>
> The permanence of the cognitive map is not a feature of the logic of communication born of a 'project' of development. It is rather a characteristic of the mechanism of development itself.
>
> (Chauveau 1992: 28)

As in the rural development that Chauveau studied, the exchanges we are examining did not affect the reproduction of development practice. However, some of the actors we observed indicated that their exchanges could move beyond such a mechanism. The time and effort that the foreigners spent in justifying why they had to act as one in front of the Chadians and vice versa underlines that this interdependence was fragile. The resentment towards a dissident foreigner (Pascal) and a dissident artisan (Osama) was expressed verbally but had no consequences for them. Had these dissidents hesitated less, they might have overcome this interdependence.

Foreign domination was more imagined than realized in Mexico, hence the necessity to resurrect the five-century-old figure of Malinche. The positions of the Mexican experts were sufficiently diverse to render the value axis and the relational axis independent. Whatever the variety of behaviours, the foreigners appeared very little in the Mexican's interpretation of the project, and only in terms of their distrust of the Mexicans. This common element was linked to the menacing Mexican Other. Despite the ability of a few local experts to interact with the foreign knowledge in a productive way (María, for example), the fear of foreign domination continued to play a role in the local reception of the foreigners. The foreigners 'bullshitted' hesitantly when they saw no other means of dealing with the Mexican engineers and, as they anticipated resistance from their local colleagues, they could find no other solution. Inflexibility in the relational attitude was more of a problem among the Mexicans than among the foreigners because the foreigners were more likely to read their own incapacity to act as a personal failure. A top international expert should always be able to adapt to his environment. Accordingly, the foreigners allowed themselves to reinforce their image of the Other on the basis of their daily experiences less than they allowed their Mexican colleagues to do so. The Mexicans, for their part, actually blamed the foreigners (the gringo) for the lack of communication that resulted from their own incapacity to react to the offer of technical knowledge.

The exchanges in Mexico were thus blocked by the cultural distance (alterity). This distance was a monument to foreign domination that was supported by the symbolic bedrock on each side of the interface. Nevertheless, the traces of cultural distance described here were the only evidence of real foreign influence in Mexico. The relationship of the foreign consulting company and the Mexican companies to the Mexican Energy Agency was of a purely commercial nature. Power resided solely with whoever was the client. Among the experts, power consisted in technical capacity and in the reputation of Hagler, Bailly, Inc., which exercised a certain cultural influence in Mexico. Contrary to the situation in Chad, the cultural distance in Mexico was not based on the rigidity of the relational axis towards the value axis (the relational attitude of the actors determined by the value of the Other) but on its relationship to the identity axis (the recognition of the identity of the other at hand).

The gringo did not enjoy the same status among all the Mexicans. Vincente, Geraldo, Aníbal and Humberto recognized the Other raised by those Mexicans who had never had the opportunity to travel or benefit from modern professional training. The foreigners were less alienated by the more informed Mexicans, who saw differences among the foreigners. For the younger local experts (Ramón, Carlos, Lorenzo, Juan, María and Eva), the image of the gringo was fixed and posed a problem. From my perspective as the foreign interviewer, the Other seemed to vary among the Mexicans, whereas in Chad the Other appeared to be a single shared construction. My personality had an effect on the formation of the Other in Mexico. John's efforts to 'not sound so foreign' were also directly received, as Carlos had explained when he spoke of John's criticism that *'makes you stupid'*.[31]

Although the Mexicans were able to distinguish between individual and personal character in the group of foreigners, this awareness could not overcome the blinding force of the Other. Even María, who succeeded best in working closely with the foreigners, did not dare to ask all the questions she wanted to ask to validate her work. She felt a hesitation that came to characterize all of her professional relationships with the foreigners. The foreigners also could not question the Mexican engineers in ways which helped them to understand the Mexicans' motivations and capacities. Thus, the possible identities of the present other were limited by the Other on each side of the interface, although the identities were never entirely destroyed.[32]

The construction of cultural distance (alterity) depended upon the structures that shaped the actors' perspective. The introduction of the three axes illuminated the logic behind the responses to the other in the cases of both the Mexican experts and the Chadian artisans. This tool helped us to make sense of their exchanges. However, it was rather the absence of exchange among the local actors themselves that shaped their relationship with the foreigners, and thus conditioned the inertia of the group as a whole. We must therefore consider these dimensions in terms of the colonial past of the local context. It seems impossible to use the complex image of the gringo, as he appeared to the Mexicans, as a way of deriving the meaning of the exchanges among them. The dimensions provide the key to understanding how the communication among the Mexican engineers might have been successful, who between John and Jim best facilitated the work of the Mexicans, and who knew best how to get what they needed from them.

The differences in the behaviours along the three axes show that the Other was not the sole motivation for the actors in Mexico. These differences underline the impossibility of doing as a group what most had begun to accomplish individually. The same remark can be made regarding the usefulness of the axes in the interpretation of the artisans' behaviour in Chad and the importance of the figure of the nasarra. The outstanding foreigners, John and Pascal, whose attitudes were consistent and strong, emerge as exceptional individuals. But the exchanges that took place between the foreigners themselves were less determined by the developer–developee encounter than were those between the local experts. The French (Jacques and Martin) worked closely together and their exchanges about Appui Technique were rich, but they shared little about their imaginary interlocutor, the poor needy African. Similarly, the US engineers did not mention their perception of the 'obstinate Mexican' to each other. On the other hand, colonialism came up in discussions among Martin, Jacques and Pascal. The French presence in Chad frequently served as the main topic of conversation over lunch.[33] Their protagonism required cultural distance, and they supported each other in the cultural distance while their concern for the developee was limited.

In summary, the behaviour of the local actors was more complex than the behaviour of the foreigners. Reducing the local perspective to the construction of the gringo or the nasarra is simply unsatisfying. In order to grasp how cultural distance (alterity) was stabilized on both sides of the interface, it is necessary to show the degrees of freedom on Todorov's three axes. Ultimately, for this study, the constructions of cultural distance provide less insight than the endo- and exo-social processes, which determined the fate of the know-how in question. In short, the communication analysed here was more structured around the exchange content than around the cultural distance. For management options, however, we have a basis for qualifying tools that change the events. Accepting the latent processes as given,[34] management concerns the reciprocity of cultural distance. The dimensions of the encounter allow the qualification of any management tool and its consistency. This is the basis upon which to extrapolate from the cases.

6.2 Conceptualizing management goals

Appui Technique and Autogeneración were not typical aid and assistance projects. They were extreme cases. Cultural distance between the experts (foreign and local) was not an issue in Autogeneración,[35] whereas, in Appui Technique, the actors were as far as they could be from mutual understanding.[36] Most TA projects fall somewhere between these extremes.[37] Therefore, an extrapolation from the implementation yields results that are pertinent for the implementation of less extraordinary ones.

We follow two assertive experts, Osama (Chad) and Ramón (Mexico), in their use of my presence as an observer. We have likened the observer role in the encounter to a little hole in a pressure cooker (section 3.2). The result is consistent with the dimensions of the encounter and with the crucial interdependencies we have just established. Therefore, we assume that appropriate management goals can be derived from the exchanges in a particular encounter. We propose that the experts can be empowered to pursue further their mutual interests.

If communication becomes an ideological issue when it is facilitated by the observer on a project, how can we talk about the conditions of this facilitation? There are several possible outcomes regarding the two particular examples. If the facilitating role of the observer aids the actors in the construction of identity, then we observe an increase in the symbolic exchanges. The production of identity allowed the Mexican experts to become comfortable with the foreigners' knowledge without the threat of domination by it (the reconquista). The likely outcome of this experience is the normalization of the knowledge in their imaginary, i.e. the removal of a foreignness that was defined in terms of the foreigner's identity as an expert. In Chad, the artisans and experts had to break free of the big brother and the roublard, or at least dissociate the technical knowledge and, thus, its otherness from them.

What can be said of the symbolic acts? The presence of the observer also influenced the logic of communication. Like a living symbol of interrogation, my involvement altered the significance of certain events and the meaning of certain statements. My presence enhanced the permeability of the interface. Habermas uses the word 'porous' to describe that permeability. What can be said of this porousness? It is part of the background (cultural preconceptions) of the actors' exchanges. This background to the communication drama consists of interpretations of the actors' interaction and the ascription of symbols to occurrences. The interface necessarily lies between two sets of symbolic orientations that can be at odds with each other. Inside these sets, the components exist in a *sui generis* relationship without causality.

There is an immediate usage of the symbolic in which the subject submits to domination by this one, but there is also a lucid thoughtful usage. Even if the latter usage can never be guaranteed a priori (we cannot construct a language nor even an algorithm in which error is mechanically impossible), it happens and it indicates the way to, and the possibility of, another relationship in which the symbolic is no longer autonomous and can be rendered appropriate to the content.[38] Even if there is no causal force behind the symbolic orientations within each set, there is still hope of finding the cause of the difference between radically divergent orientations, such as that occurring around the interface. The interface's character is arbitrary except for the personalities of the actors that individually help to shape it. For the foreigner, the nature of the sympathetic African and the consultant (Appui Technique) or the nature of national pride and world sense (Autogeneración) has nothing to do with cause or logic. The porousness of the interface is the result of the confrontation of the subjectivities of the actors. If the lucid usage is possible, it heightens the appropriateness of the symbolic to the content. But can an actor render his/her symbolic orientations more appropriate to his/her experience? The conscious usage is preferable because, as was witnessed in the two case studies, the inappropriateness of their images was so obvious and oppressive to the actors that they continually second-guessed themselves. But lucidity is threatened by the fog of arbitrariness that characterizes the interface and prevents the actors from seeing how to tailor the symbolic to the content. Their capacity to tailor the symbolic is the prime target; whether their tailoring moves in the right direction is a question we neglect.

It is impossible to package the technical object for export and send it with instructions for use, whether it be as complex as a computer-generated model or as simple as the design for an oxcart. It is accompanied by context-specific know-how which must be introduced properly – or, at least, by an expert who is theoretically capable of doing so. Furthermore, the object can never be so specialized that its use is determined. It can only ever be tailored for an individual extrinsic mode of employment. One must always modify computer systems and interpret their results in the hopes of refining their capabilities, and experts must determine the most efficient use of the turbine for the generation of electricity. The confusion is not diminished by a reduction in the presence of the foreigner, who supplies at least a basic know-how that he/she has acquired in advance.

The acceptance of the foreign presence depends upon his/her foreignness. Is his/her presence alone sufficient to temper the strangeness of the object and to reinforce local identity? The answer to this is, in fact, 'No'. Dramaturgical and communicational acts buoy up the necessary symbolic work that produces these results. Reinforcing clear and conscious action and working towards appropriateness are the first steps in realizing acceptance.

Concerning myself, the processes of symbolic adaptation and acceptance were effective. My participation in Autogeneración was judged to be a great success. The management in Mexico and in Washington, DC, agreed that I had saved the project – the technical results that I compiled constituted half of the total work. All the other experts had much more experience in the field than I and had worked on the project for just as long. Nevertheless, I alone continually interrogated the nature of the relationship between the foreigners and the Mexican engineers. The others needed my work to raise these issues for them.

One possible solution to the problem of acceptance is for experts in TA projects to carry out ethnographic work.[39] But such work is at odds with their technical training. How can we change the bond between foreigners and local actors in order to promote communicational and dramaturgical acts and discourage strategic acts?

The particular circumstances of these projects favoured success because the teams were composed of an equal number of foreigners and local actors. Judging them all according to the same performance criteria and accepting the results of the actors' work as a group effort were ways of pressuring the experts to consider the appropriateness of their orientation towards the other. This pressure was greater still when we put professionally compatible foreigners and local actors in working teams of two. But there was only limited hope of overcoming the experts' obstacles to communication on the level of global organization. After much experience and reflection in many countries, some development agencies have integrated these criteria into their project planning (Fry and Thurber 1989: 7).

But organizational solutions for the problematic acceptance of the developer are limited by the nature of the obstacles facing developers in the field. The local actors find themselves caught in a structure that blocks the expression of the symbolic domination. This structure is tainted at its core and by the general public's demand for aid in the Third World to end world suffering.[40] But the developers who are in the field, who come with the money and attempt to forge partnerships all the while knowing that the monetary support of the development effort is paternalistic, struggle most. No organizational change can hide the contradiction that developers and developees live in the field. Management goals for implementation have to be derived from each developer–developee encounter. The following demonstration is the first task and constitutes an unprecedented approach to TA implementation.

The opposition between the exo- and endo-social processes (section 5.2) provides the initial task for management. In an exo-social context, the technical object is to be preserved, so that the potential of the know-how survives the encounter, whereas the endo-social process makes the know-how more visible in the exchange and, consequently, less vulnerable. In an endo-social context, the task is thus to facilitate the adaptation of know-how to the context. The know-how should be more actively negotiated between the experts.

This difference is manifest in the actors' behaviour. Allowing the object to exist independent of the construction of cultural distance helped Osama to operate like Mohammad and helped Jacques to relate to the artisans as Pascal did. Such a change in behaviour was not suitable for Autogeneración because the separation between identity and know-how resulted in Miguel's appreciation of inexperienced young engineers (like me), based solely on their foreignness. Both Ramón and Miguel were finally able to move beyond their disagreement with the 'colonist' over the computer-generated models of the thermodynamics of turbines. Before specifying the initial task in Autogeneración, we assess how the Chadian artisans could be helped to pursue the initial task in Appui Technique.

In order to understand how the technical object can survive the construction of cultural distance (alterity) in an exo-social context such as Appui Technique, we must clarify two options:

1   Whether cultural distance can be generated differently.
2   Whether the construction of distance must necessarily be reduced for the know-how to survive.

During the course of Appui Technique, it often seemed that the technical objects, such as the quality of the manual work, were equally accessible to, and appreciated by, the Chadian artisans and experts alike. Both groups often used technical reasoning to advance discussions. Rahman (the artisan who rejected a relationship of cultural distance from the experts) was fascinated by Pascal's 'secret' knowledge of trigonometry. The introduction of totally new knowledge was read by the artisans as a sign of foreign understanding hiding below the surface of discussions, and that could complicate things. Secrets make transactions difficult. But Rahman and Mohammad integrated all that they knew into their work without hesitation.

Rahman saw the many ways in which France differed from Chad and, therefore, felt no need to reproduce this distance in his professional relationships. He was able to work very closely with the foreigners. He was forced to reflect critically on the knowledge he was given knowing that his performance ('informed by the White man') would be judged harshly. In this sense, he had good reason to want to persuade the experts to consider long-term commitment to the artisans. Martin was often surprised by Rahman's candour. It is clear from this example that Rahman could construct cultural distance in relation to the foreigners and independent of the technical objects. Other artisans seemed simply to wonder how any foreign technical objects could be adaptable or accessible to them. The following exchange, which took place during a break while working on the oxcart prototypes, illustrates their belief that they could not achieve the same standard of workmanship as the French:

Ngerbo: It's not like in France, it doesn't go this far. [Looking bitterly at a finished piece.]
Rahman: But what about in Niger, huh? You have to stay close by, you can't go too far.
Ngerbo: But isn't that already Cameroon?
Dambai: In the Sudan, you know ... they are hard-working.

It seems that Rahman was initially more optimistic about the acquisition of technical knowledge than the other artisans. In any case, cultural distance from the foreigners could have been separated from the technical knowledge if this was not already the case. Although French standards were beyond reach, the Chadian artisans could reach the same standard of quality as in other countries with hard-working artisans. Once this separation was achieved, the artisans found it easier to reply. Their discussions often intensified more rapidly when the work was advancing well. We can thus maintain that it was possible in the context of an exo-social process to construct cultural distance in a different way. To protect know-how, the actors in TA projects must consider ways to put it aside, to reduce its importance in the relationship between the two sides of the interface.

To look more closely at this alternative mode of construction, it is useful to consider the case of Osama, the artisan who constructed cultural distance most vehemently. He was the only artisan who had travelled in Europe, and he returned home full of admiration for what he had seen. In his interview, he cited Le Métro and other advancements in transportation as projects that set the standards for the future. And yet, Osama lived in his own private world, ignoring the foreign knowledge in the workshop. As we have seen, his vision of cultural distance was determined from the outside, not generated from within the project experience.

Osama could not express his criticism to the experts directly, so he expressed it through his judgement of the technical objects. He would often dramatize the treatment of the other artisans by the 'colonist' in front of the foreign experts, claiming that the colonist engaged in a kind of psychological domination, exaggerating his faults – his way of damaging and of breaking drill bits.[41] Domination was partially a self-imposed condition, as artisans like Osama did not respond to the foreigners' attempts to dialogue about the technical objects.

During his interview, the technical objects lost some of their cultural distance. Osama was able to think more objectively about them because the interview allowed him to question his image of the Other. The next day, he seemed critical of local knowledge in a way that he had never been before. At the beginning of his interview, Osama wanted to learn how to use a tape recorder, specifically to know how to stop it so that he could control the recording of his interview.[42] Several moments of real connection between us prompted Osama's admission that he saw me '*as a student on the project, like the artisans also*'. He seemed genuinely shocked by the relative similarity of our judgements of the collaboration. '*Yes, yes, you don't know anything about that … now I am not afraid to go to Europe … It's like you trying to learn Arabic*'. On the other hand, when I mentioned GRET, the financial end of the project (the IBRD) or the official responsibilities of Martin and Tahem, Osama fell silent and refused to discuss the issues.[43] Martin and I were a different sort of nasarra for Osama because we did not drive cars. This perspective led him to denounce the project as a deceptive show of assistance and the Chadians as fools '*who *allowed the nasarras to play these games'.*[44] He favoured talking much more about the technical issues involved in the project. The foreigners became more and more human to him as time went on and he began to wonder about their personal areas of expertise and their individual commitment to the project (when and why they would leave, etc.). He also pointed out the artisans' strengths, showed how the organization of labour in the production of prototypes was sound and how they could help each other productively. In this way, the technical objects had a much lower symbolic charge as time went on. *'Personally, I don't know any other way to do it. This is the best way, that's it'.* He went so far as to admit that Appui Technique responded to a local need and could help the artisans advance technically and economically in a way that no other project had before. His way of negotiating the project experience would certainly have improved foreigner–local relations in time had it not been for Osama's poor performance as a representative of the group of artisans as a whole.

Osama began to feel an affinity for me of his own accord, without any initiative on my part. He was simply interested in my perspective on the project. In fact, he was most affected by my personal situation – for example, I spoke only a few words of Arabic and drove a moped to work instead of a car.[45] My presence in the communication field seemed to facilitate his use of the interview for his symbolic work. His reflection soon became a stronger influence on his thought than the power differential between the civilizing colonial and the native.[46] Symbolic work had different motivations for the other actors and was often informed by a lived experience of collaboration rather than contradiction, or the nature of an actor's trade.

Osama's reaction in the interview confirms the initial task: only the introduction of other knowledge capable of replacing the knowledge of the foreigners as an expression of cultural distance can augment the permeability of the interface in an exo-social space such as Appui Technique. It further reveals that this other knowledge must be present during the construction of the interface and must be linked to a foreigner; in the case of Osama, to me.[47] This statement summarizes what we can learn about the adaptability of the interface based on the analysis of the researcher's position in the field. We can add a word concerning what we know about the effect of the technical objects (the knowledge itself) as well. In Appui Technique, all technical knowledge could be emptied of its substance and become a source of resistance in technical discussions. This resistance could be neither eliminated nor reduced by the actors despite all the moral claims of contributing to economic growth to which they had recourse. It was thus impossible to affect the relations between the French and the Chadians by altering the technical content of the knowledge in Appui Technique. This operation is exactly the opposite of trying to adapt the relationships to the encounter. Similarly, in Chad the encounter was greatly influenced by the relationships among the Chadians. Other artisans followed the lead of Rahman, Osama and Mohammad, but for their efforts to have truly disabled the interface they would have had to act as a united front.

If other objects must be introduced as markers of cultural distance in order to liberate the technical knowledge from its cultural fetters, these objects must be part of the encounter and shared among the local actors. Furthermore, the technical object must be replaced as a symbol of cultural distance without losing the development objectives of a TA project. Such a shift of cultural distance was achieved by Mohammad and Rahman with ease and by Osama through intense efforts.[48] The management goal in Appui Technique was to enable all actors to realize such a shift of cultural distance. Osama's endeavour allowed him to see the preconditions of this goal. Despite the necessity of these preconditions, we cannot know whether these are sufficient for such a shift in other developer–developee encounters.

We now turn to Autogeneración, the endo-social case. In Mexico, the local actors did not endeavour to mark the difference of the foreigner, and never tried to exploit technical objects to affirm Mexican identity. Their symbolic work was restricted to the encounter itself, where they sought confirmation of the local culture in the recognition of their knowledge. Ramón's interview will again serve as a gauge as he was most earnest in his attempts to collaborate with the US engineers. It was he who quipped in response to my request for an interview, '*I hate the gringos, but I love their money*'. During the course of the interview, he explained that the technical content reflected the incapacity of the Mexican authorities to accept the fact that Mexican engineers are as capable as foreign ones of achieving modernization. The whole situation was a farce, he explained, that some acted better than others:

Ramón: It is sometimes hard for me not to see them as larger. I have always had respect for people who know what they are talking about, but I try to surpass them. All of us respect John and you for what you know but that doesn't mean that you are better than us. That's the difference. On the other hand, Juan and Miguel think you are better in everything – they are crazy. One day I'll say to John, 'You're wrong, you made a mistake, it's incorrect!' Because here people don't see that. They think, how could you say that to a gringo? Damn, that annoys me. [Interview.]

The next day, when the experts met in the office, Ramón began to denounce the foreigners' pretentiousness ('Superman', see p. 109). He had not used his interview to reflect on his relationship with the imaginary during his intercourse with the US engineers. Rather, by expounding upon his professional identity for 4 hours in front of a foreigner, he effectively reaffirmed it. In order to break free of his submission to the foreigner, he set his sights on the Other and prepared to surpass him in skill and innovation. In fact, Ramón contributed to the preparation of an engineering manual and personally presented the results to the Energy Agency, but John thought he could have assumed even more responsibility.

The interview gave Ramón an opportunity to denounce the gringo, and after it he made more use of 'foreign' technical knowledge in his work. During conversation with an observer, he was able to push his confrontation with the gringo further, primarily because the other experts had publicly used me in this capacity and also because I explicitly agreed to entertain the farce of 'serving the Energy Agency' produced by Autogeneración. No other foreign expert helped Ramón to negotiate the Other. As the project went on, Ramón worked more closely with John (who learnt his trade in the USA) and then with Jack (who learnt his trade in France). Both of them concluded from their interaction with him, which might be characterized as difficult, that he was not at all motivated in his collaboration with the foreigners. Ramón did attempt to work with the foreigners as equals, but apparently lacked the subtlety to adapt to the habits of individual engineers – even two engineers with very different ways of working.[49]

We have shown that the gringo represented a real threat to the Mexican experts. Ramón confronted this threat head-on. In order to accomplish this feat of resistance, he had to create opportunities to affirm his identity through, and in terms of, the technical content; the content that gave meaning to the project, its pertinence and the various aspects of its practical development. José even managed to affirm his identity through this process: '*We learned who we are*', he declared in different interviews with me. His vast experience with the material and as a technocrat gave him a sense of his own work that Ramón failed to develop, concerned as he was with the gringo rather than with his engineering prowess. The very act of recognizing the valid contribution of a local expert set a precedent for the appreciation of the Mexican team as a whole.

The local experts in Autogeneración had to create media (or vehicles) that would distinguish their results as specifically Mexican. That such a medium would have to be visible and of an non-ambivalent nature was clear in the statements made by Juan in his interviews. He explained that the results reflected neither the abilities of the US engineers nor the competence of the Mexican engineers. And yet given the collaborative nature of the project, no one would ever have suggested separating the technical analyses by nationality into those prepared by the foreigners and those prepared by the Mexicans. A separation of the experts' labour would have signified a rejection of the universality of analysis. Thus, any medium of distinction would also have to avoid the qualitative differentiation of foreign and Mexican tasks.

Doing this would necessarily distribute the engineering tasks between foreigners and Mexicans, allowing for foreign and Mexican influences, which would enable the Mexicans to measure themselves against the foreigners (important to Miguel and Ramón). However, this medium would also have to be able to accommodate technical know-how in order to proceed in anticipation of the differences which would arise throughout the process of identification (important to María and José). Given the Energy Agency's preference for foreign knowledge, it would be imperative that the medium remain within the purview of the experts, protected from possible appropriation by the outside. The experts needed context specificity at their discretion – to be able to state that this know-how is correct under these circumstances and that with other circumstances different know-how applies. This would have enabled Mexican experts such as Ramón and Miguel to affirm their professionalism as José was able to. The management goal in Autogeneración was to enable all experts to create context specificity of know-how.

In both case studies, the management goals contained nothing that had not already occurred in these projects. These goals focused management on the implementation obstacles of the technological objectives. The medium that would enable the Mexicans to demonstrate their abilities and the shift that would allow the Chadians to mark their cultural distance (alterity) against the foreigners have, in common, their dependence upon the cohesion of the local actors. The relationships among the Chadians were nourished by external factors (exo-social), whereas the relationships among the Mexicans grew out of internal conditions (endo-social). However, in both cases, the reinforcement of these relations allowed the local actors to resist and, hence, reduce the cultural force of the foreigners. For their part, the foreigners could only orient their approach to the extent that the local actors were willing to respond specifically to each of their individual capacities.[50] In order for the foreigners to move beyond their construction of the pride of the Mexican and of the Chadian 'roublard', he/she would have to see his/her contribution to the project validated by a local expert. In short, the relationship between the knowledges and the relationships among the experts are co-extensive and are equally affected by changes in the encounter.

Signifiers of difference between the local actors and me permitted Osama to dissociate the technical knowledge from the symbol of the nasarra. Foreign knowledge (the technical object) could only be liberated from its origin when its roots could be displaced and replanted in other soil, i.e. the label of the object took precedence over its content in Chad.

The difference between the lived experiences of Ramón and that of José in Mexico turned rather on the independence of the label and the content of the object. José accumulated identity by making use of the technical content, whereas Ramón dissipated local identity. Although it may appear reductive to characterize the lived experience of these actors in this way, the reduction highlights a difference in perspective that can help us to understand the consequences of an encounter.

We can thus discern two types of identity construction at work in TA projects. In the Chadian case, the actors weakened the link between the origin and the content of the knowledge because its politically charged origin blocked access to the knowledge in itself. In the Mexican case, they strengthened the link of the origin and the content because the content was meaningless without that link. The basic distinction between these types is not between strengthening or reducing that link. Reducing and reinforcing the link between the origin and the content of the object are gestures that mask a deeper cultural conflict: the foreignness of the origin and the local reading of the content.

Friedman identifies cultural processes within larger global systems that shed light on identity constructions. TA events are ideal for studying cultural processes because their interfaces are sensitive to transformations that constitute and regulate identity. In order to demonstrate the complementarity of the foreign label (Chad) and the local meaning of the content (Mexico), we introduced the tenets of global anthropology through Friedman's work: the contents where the formation of identity takes place constitute a potential global space that circumscribes identity, a panorama which includes the interaction between local identity formation and the dynamics of positions in the global space (Friedman 1992a). Friedman shows the political importance of the nature of identity formation and its negotiation in the social imaginary.

In the case studies, the figures of the gringo and the nasarra were autochthonous, born of a sociopolitical history. The foreign experts did not resemble these figures, but the local actors positioned themselves against the former in terms of a global imaginary that was nonetheless constructed locally. Within this identity construction, the circulation of knowledge took place. Thus, this circulation remains marked by its local origin. The complementarity of the two types of symbolic work lies in their shared function of positioning the local actors within the locally constituted global space.

A second indication of the complementarity concerns the efforts of Osama and Ramón and the structures of communication. Ramón defined the Other in negative terms on the value axis (as did many of the Mexicans) as well as the relational axis, which led to his confrontation with the foreigners. Ramón skilfully used the modes of dramaturgical communication (much more often than the other Mexicans) and considered acceptance of the strategic mode to be a personal failure. Osama showed less respect for the nasarra than the other Chadian artisans, nonetheless he rejected more vehemently the possibility of sharing their knowledge. Among the artisans, he saw the fewest risks in the dramaturgical modes. He strongly resented the resignation to the strategic mode, but saw in every contact with the French an invitation to return to that mode.[51] Based on this evidence, we can conclude that the complementary nature of the two types of symbolic work corresponds to the structures of communication. Osama and Ramón both had recourse to dramaturgical acts when the opportunity presented itself, but their failures were different. Osama's failure was directed towards himself and Ramón's was projected outward.

If we can study the historical positioning that actors construct to interpret their relations, it is because this positioning is neither assured nor obvious. It reflects the individual efforts of an actor within the limits of historical conditions. With each new encounter the positioning is unique, but the very act of positioning oneself creates the possibility of constructing knowledge from this perspective. The differences in the structures of communication show that, independent of historical context, the efforts on the part of the foreign and the local actors to establish a reciprocal rapport are a function of the different types of exchanges between the groups of actors. The transformation of cultural distance into the protagonism of the French and the Chadians' defensive resistance to the Other in N'Djaména, and the interplay between the ethic of the encounter and the modernizing force of the project in Mexico City, can be read interdependently in this way.

6.2.1 Conditions for applied research

By tracing the development of the work of the actors, the analysis of ethnographic data brings about the positioning of the foreigners and the local actors, one that generates the possibility of restitution by producing diverse and rich reactions to the results. We began with an interrogation of the nature of TA projects. The question soon evolved into a suspicion of the validity of the ethnological perspective. The presence of structural contradictions – expressed here in concrete terms as the difference between Osama and Ramón – led us to much deeper issues. Because foreign experts share an interest in the more 'exotic' objects of ethnology,[52] my experience with them led us to ponder the conditions of knowledge about the Other.

Since its inception, anthropology has accepted the link between the possibility of knowledge and colonial power. Both the ethnologist and the developer seek in their own ways to constitute knowledge around the development aid.[53] But this doubling is not symmetrical and these developers do not have the same capacities or positions in the field. By demonstrating the reciprocity of cultural distance in Appui Technique, we indicated that structural blockage was not the defining principle of this project. The analysis of the actors' exchanges permitted us to ethnologize further. The day-to-day experience in TA is more flexible than researchers believe it to be and more rigid than the TA professionals will admit. The result is that the practicalities forgotten by the experts always resurface and the theoretical models that the researchers construct are always inadequate.

The conditions in the field enable researchers to work. The analytic frame that permits the reconstruction of cultural distance raises new questions because it forces the experts to negotiate their knowledge. Such a negotiation can become a new epistemological niche for anthropology. The topical relevance of this fieldwork consists in the contemporaneousness of the actors involved, in the historical circumstances in which they confront the past and, finally, in the possibility of constructing identity without relying on the differences across the interface, i.e. without reinforcing cultural distance. Today, such fieldwork is possible and necessary and need not imply questions about the legitimacy and the authority of the anthropologist. Projects that involve the direct confrontation of warring types of knowledge offer research the opportunity to find new angles for analysis. This book contributes in precisely this way to the advancement of the anthropology of development.

6.3   Implementing technical assistance

One management goal is to shift the construction of cultural distance away from the technical content towards other elements of the encounter which are linked to the foreigners and shared among the local actors. The other goal is to enable the experts to create context specificity of know-how and mark their contribution as Mexican. The broader objective is to reveal the interdependencies of cultural distance, in the first case, by separating the value axis from the relational axis, and, in the second case, by improving the recognition of the other at hand (the identity axis). We excluded the content of TA. Thus, we suggest that the planning incorporated in TA events based on the TA content is not problematic, it is technically and economically sound (which it actually is not).

What is left is operational management – the day-to-day decision-making. None of the following management tools have intrinsic (sufficient) properties which would affect a developer–developee encounter. But, in a specific encounter, they contribute to achieving these goals. Most of these tools are not currently used as such. While we will not assess them separately here, each one should be adapted to specific TA events and completed with corresponding measures.

Tools to shift cultural distance away from the technical objects:

- Differentiation of non-essential aspects related to foreign and local participants and of personal concerns such as working hours, clothing, transport and food.
- Establishment of dictionary for all technical terms, acknowledgement of equivalence at all occasions, addition of vocabulary suggestions from all participants.
- Separate meetings of local and foreign experts with elaboration of a common agenda for both, while accepting only combined reports as official project documents.
- Extensive data gathering, data administration and elaboration, and making the results widely available.
- Horizontal structure of tasks, where foreign output is also local input and vice versa.
- Integrated documentation of expert performance and other reporting arrangements.
- Housekeeping and inventories maintained by local experts.
- Budgeting and milestones in implementation defined as simply as possible, ideally with standardized parameters maintained from the beginning to the end of the project.
- Defining simple quality-control parameters recurrently and in writing, distributing auxiliary data and intermediate calculations.
- More line positions than staff positions.
- Absence of references to specific contributions from foreigners or local experts in results and products.

Tools to mark the context specificity of know-how and especially its local origin:

- Non-essential aspects of participant conditions varied individually.
- Project products different for each group when suitable.
- Organizational differences marked relative to objects, specific schedules for different applications or examples.
- Whenever possible do not use the first language of the participants, avoid use of metaphors.
- Separate meetings for foreign and local experts with each group documenting the changing agenda over the project period, some of these documents becoming official project documents.
- Data gathering and administration initiated at discretion and as non-standard.
- Vertical structure of tasks where foreigners perform one application and local experts another; tasks are chosen based on differences in skills needed and those available among the experts.
- Emphasis on informal communication between experts, sharing of resources with these remaining connected to individual experts.
- Acknowledgement of the importance of tacit knowledge in expert competence.
- Resolving role conflicts among the experts through requiring compromise, yet not avoiding competition.
- Designating housekeeping specifically for each task, not as a general role.
- Keeping budgeting separate but shifting resources when tasks are moved between foreign and local experts.
- Devising an implementation diagram in a 'public' location, such as a meeting room or entrance hall, where all experts can make adjustments and mark completion.
- Results and reporting arrangements with specified contributions from an expert.

These tools are instructive examples of the differences between the two lists. In other words, it is important how these management instruments constitute different categories, corresponding to particular conjunctures of developer–developee encounters. Each list could be much longer, but their specificities are in themselves a management principle. This specificity is a class of management knowledge, below the level of the project unit. Most management tools used in TA practice can be shaped to serve the first management goal or, alternatively, to serve the second. It is important to recall that these management goals are encounter specific and have to be redefined for other encounters. The lists are precedents for such definitions. In this chapter, we review more evidence for such management goals.

The projection (or thought experiment, Table 5.1) of the three latent processes on evaluations showed how these appear in TA evaluation documents. We cannot proceed similarly with TA management. While evaluations are abundant (estimates go up to 200,000 evaluation documents produced so far), the literature on TA management is scarce, essentially limited to a handful of documents.

The dismal state of knowledge about TA management has been apparent since the 1960s.[54] Most of the reform attempts have not been taken up because of the institutional and political rigidities. But the impression that all insights behind the reforms are old and just repeat themselves is certainly mistaken. Reforms have indeed taken place.[55] The necessary comprehensive review cannot be elaborated here. The similarity between some of our insights and previous reforms does not avoid the possibility of a renewed critique, with new evidence, having an impact on TA management reform, particularly as we started from development anthropology and proceeded step by step to management. Therefore, we proceed with selective comments:

> Thus far, AID staff and research contractors have used rather vague definitions of management performance improvement that may be so broad as to be meaningless, either for their own research or for formulating strategies of intervention in other societies and cultures.
>
> (Rondinelli 1987: 155)

In the past, TA management went through a succession of orientations. Often, these were just shadows of management fads coming out of business administration. 'Performance management' in the early 1980s, 'strategic management' in the 1990s, organizational development and, presently, organizational learning.[56] These management approaches arise in industrial corporations. Those appearing within development agencies, as far as they concern management approaches to TA practice, are not adopted, according to Rondinelli and many others (Rondinelli 1987),[57] primarily for political reasons. Rondinelli stressed in particular the need for control over aid resources as a key factor for the rigidity of management habits. Indeed, Forss *et al.* found that a large part of TA consisted of control functions, however they are concealed in project papers.

Forss asserts that much of the additional insight on TA over the last 10 years concerns new roles for foreign experts:

> Even when broad agreement has been obtained on priorities, it appears at times that technical assistance is virtually 'thrown' at problems, without a careful analysis of the feasibility of attaining the proposed objectives, or of the approaches and methods that would be needed to achieve them.
>
> (IADB 1988: 19)

The blame for implementation difficulties could easily be assigned to all the parties involved – the Bank, the borrower/beneficiary, and the implementing personnel – but, above all, it lies in the absence of well defined roles for each of them. This becomes dramatically apparent during the crucial phase of implementation, i.e. project start-up.

(ibid.: 23)[58]

Whereas the presumed functions of the foreigner were those of an advisor and a trainer, Forss's own evaluation revealed the gap-filler, the institution-builder and the gate-keeper/controller roles. These roles have been confirmed by some researchers, whereas others suggest, for example, process consultant, mobilizer, doctor and performer. The attention to the possible roles of foreign experts is not related to the implicitness of the control functions in project documents. Possibly, this attention is only an expression of the confidence in intrinsic qualities of modern technological knowledge, underestimating the know-how necessary to transmit and apply such knowledge. Therefore, Hirschman's 30-year-old observations on IBRD projects are still important and are yet to be translated into a sufficient uncoupling of the project cycle definitions in TA management (Hirschman 1967, 1977).

Development agencies are experimenting with these roles and the corresponding consequences for TA management. These experiments in management have two deficiencies. First, these roles concern mainly the foreigner. What is missing from the research are respective roles for local experts. The ruin of the expert counterpart model could well have been that there was no scope for different types of counterparts in TA management. Second, as the case studies have abundantly shown, these roles are the result of each particular developer–developee encounter. Whatever role concept is used in TA planning and management, it is a strait-jacket for the TA actors and impoverishes their encounter. While role definitions are necessary for planning and evaluation, they are considerably less useful during implementation. One way of improving TA management fundamentally is to enable local and foreign participants to swap roles at their convenience and to enable them to ignore whatever role concept was used during planning.

Local and foreign experts who confidently ignore the roles anticipated during planning can concentrate on exchanging knowledge. Management can facilitate local and foreign experts' experiments to invigorate their encounter. The success of participatory approaches depends on this and will continue to do so. It is tempting to translate appropriate changes of the constructions of cultural distance (according to Todorov's dimensions) into roles which facilitate their collaboration. Much of the literature on TA management does precisely this in an implicit manner. But that makes it a vain intellectual effort as far as TA management is concerned. An important lesson to be learned is assuring every TA participant that evaluation will never be based on the roles anticipated during planning. Local experts can be forced as much as foreigners are to change roles. One possibility is to define several roles for both foreign and local participants in the design of TA.

A major argument appears frequently in the literature concerning TA: since the implementation of these projects is so forceful, better preparation makes little difference.

Only improving management tools during implementation can achieve substantial improvements. Many development agencies accept that project objectives evolve and the fact that the original objectives were not realized is no indication of failure. The challenge is how to devise implementation as a flexible process that allows the objectives to evolve (Farrington *et al.* 1998). TA management as a flexible process should eventually reflect the complexity of the social processes which are affected by TA (of which the latent processes are only the most salient ones).

The IBRD's *Handbook on TA* (IBRD 1993) is a unique reference for comment on current management reform. This is not because the case studies were IBRD funded, but because of the IBRD's authority and its considerable information and research advance over all other agencies. The Handbook was meant to be an in-progress publication, but it acquired the influence of a standard book, providing guidance to IBRD task managers on how to design and manage TA.

The principle comment to be made about the Handbook is its omission of the abovementioned tools. While its authors continuously stress the need to improve the IBRD's influence on day-to-day management, the Handbook lacks the concepts or variables to put day-to-day issues on the agenda of appraisal, design, supervision or evaluation. Possibly, the ultimate purpose of the Handbook is to enable IBRD's task managers to avoid some of their 'hidden rationalities' by addressing day-to-day management issues directly and also during negotiations with loan recipients. This would explain why the management tools do not appear, as they do not correspond to the presumed hidden rationalities of planners and managers. Nonetheless, these management tools can help task managers to achieve this.

Besides foreign consultants and counterparts, who are the recipients of expertise and know-how, the Handbook only refers to local experts providing 'appropriate consulting technology'. How this expertise and know-how relate to project objectives and how their realization can be organized does not appear. Among these three features there is a void. Because of the absence of this information, the TA management that is proposed contains such a degree of contradiction that its reform orientation cannot be advanced through the use of the Handbook. On the other hand, the Handbook makes powerful advances regarding the importance of institutional analysis, the place of NGOs and local consultants, and twinning arrangements in TA. These advances also express the difficulty of improving the role definitions for the consultants and other experts:

It should be noted that the TA which is dealt with in this Handbook pertains to assistance provided to public sector institutions for policy making, economic or environmental management or for investment management such as infrastructure, manufacturing or service industries.

(ibid.: v)

The selection of TA modes should consider the ability of the recipient to ensure the use of the end products and, in the process, become more self-reliant. Simpler designs with modest achievable goals are required. Be wary of becoming overly optimistic when (a) anticipating the lead time necessary to get a task started, (b) projecting the borrower's recurrent revenues, (c) hearing commitment when talking only to technocrats, and (d) claiming long-term in capacity gains.

(ibid.: 9–10)

An adaptive approach toward project design requires close monitoring of the evolution of the project as well as intensive involvement of Bank staff in the continuous process of discussions and decisions on the allocation of project funds and targets to be achieved. And, for an institution like the Bank, where traditionally twice or three times as much time is devoted to appraisal as to supervision, this is a vexing issue.

This does not imply, however, that Bank-financed IDTA [institutional development technical assistance] is incompatible with pragmatism and adaptability. The use of the process approach, in fact, would be consistent with Bank efforts to implement many of the recommendations of the Portfolio Management Task Force, whose conclusions are that (a) ..., (b) ..., (c) the project-by-project approach to portfolio performance management needs to proceed within a country context to address generic problems of implementation and systemic opportunities for portfolio improvements, and to focus accountability within the Bank for portfolio results, (d) ...

(ibid.: 63–64)

Generic implementation problems are to be dealt with on the sector or country level. This reading of the IBRD's TA practice is unsatisfactory because generic problems cannot only be addressed above the project unit on a broader scale, but ultimately these need to be addressed below the project unit, i.e. aspects of individual projects have to be addressed instead of aspects that are common to all TA within a sector. The conditions for an exchange of know-how revealed by this study have shown this abundantly.

The only project-specific (or 'subproject' level) management tool mentioned concerns the combination of long-term and short-term consultants. A suitable combination minimizes the disadvantages of each. This combination appears to be a mark of IBRD TA that distinguishes it from other donors and that opposes the rejection of gap-filling TA,[59] which the Handbook identifies in the United Nations (UN) agencies:

The long-term advisers provide back-up support for implementing ID [institutional development] improvements, and the short-term consultants conceive and develop the methodological inputs. In this framework, emphasis is placed more heavily on the 'behavioral skills' of the long-term experts than on their substantive technical knowledge.

(ibid.: 93)

The distinction between behavioural and substantive contributions corresponds to the difference between know-how (context specific, tacit) and abstract technical knowledge. The suggested link between this distinction and the time frame is, however, rather minimal. Any short-term expert mobilizes a behavioural contribution with every bit of substantive knowledge. The distinction plays a central role in TA, but it has to be accounted for in TA management in many other ways, not by the time span of expert involvement.

Two other aspects are missing from the Handbook. First, the possibility that consultants' contributions need to be adapted to such an extent that new knowledge is created does not appear. The appropriateness of TA is simply implicit and certain. Of course, it is necessary to leave the TA content out of management guidelines (as we have done as well), but this does not at all suggest that the TA content is as development relevant as developers used to believe – and, more importantly, as relevant as the developees tend to believe. Second, the possibility that foreign experts contribute something and local experts contribute something else is not denied, but that the impact of TA can be achieved only if these two contributions correspond to each other is not addressed. These are the limits in the comprehensiveness and the contribution that the Handbook makes to TA practice.

If TA crucially depends on accounting for the contributions from local and foreign participants in all their complexity, the reductionism in treating local participants in all TA encounters as generic 'counterparts' should be reflected in other parts of the Handbook. Inconsistencies and odd conclusions should appear which express the hidden complexity. Indeed, the guidelines on counterpart staff show a remarkable degree of explicit contradiction:

As a concept [counterparts] it has proved remarkably robust in the face of almost unrelieved failure. Who can recall successful examples of counterparts working alongside expatriate experts, and smoothly taking over when they leave? Is it not time to discard the traditional counterpart concept, and inject fresh thinking into this area?

(ibid.: 104)

The re-examination proposed contains only a general call for a true partnership with defined functions and responsibilities.[60] Despite these doubts and no solution, the Handbook instructs IBRD managers to treat any shortcoming very severely, and, when qualified counterparts are not assigned as planned, it has to be considered a violation of the Loan Agreement.[61] Furthermore:

> At periodic intervals, project managers should review progress in the acquisition of skills and knowledge required for each counterpart position. If the transfer of expertise to existing counterparts is below the minimally acceptable levels, new counterparts should not be brought in until the previous targets have been met.
>
> (ibid.: 108)[62]

The only conceptual help proposed consists of sample checklists to be used: each task or skill to be acquired appears with the counterpart performance to be rated from 1 to 5, and each rating has to be initialled and dated. The presumed role of the counterpart is a strait-jacket which thwarts efforts to address the complexity of the exchanges of knowledge. The sanctions concerning disbursement have to be brutal to maintain that facade. The Handbook's insistence on the counterpart functions, while pointing to their weaknesses, is evidence for the crucial rationale of the concept.

The IBRD's increased attention to institutional development reflects the previous engineering-related TA and its blueprint management approach. Institutional development TA 'requires a flexible approach, a readiness to face reverses, and modify objectives and implementation modalities, and an extended association with the borrower' (ibid.: 187). Many of the deficiencies of investment-related TA could have been avoided if these discoveries had come earlier. But this cannot be proven. What can be shown, however, is that much of the TA criticism produced in the 1970s and 1980s now reappears in institutional development approaches. However, demonstrating some causality would be a powerful contribution because it would clarify the institution-related TA objectives. The two projects studied are helpful here because they were targeted at institutional capacities in the respective sector and their impact was nonetheless limited primarily by the absence of institutional analysis and its consideration for managing implementation.

If we assume that the two projects studied represent what the Handbook, published when both were about to close, is supposed to change, then we can comment on the Handbook by assessing how its suggestions might have altered them. This of course ignores the fact that other factors prevent what the Handbook seeks: increased influence of IBRD managers on TA management. These other factors hold sway because this increase is gradual and depends on the IBRD managers' capacity and willingness to achieve it.

The insistence on the counterpart functions reinforces the first latent process: the mutual appreciation of technology. Establishing a catalogue of tasks that the counterparts are to acquire, and verifying their performance on each one, would make it more difficult to work towards the instrumental core of technical knowledge. The latent process would be reinforced (Figure 5.1) because the actors' insistence on the unjustifiable claim that ends be part of the instrumental core would be amplified. We have already alluded to the constructive approaches to technology pursued by sociologists and historians,[63] which are based on the notion of a seamless web between culture and technology. Because of this seamless web, cataloguing and verifying counterparts' tasks will not improve their co-operation with foreigners. In the medium term, the insistence on the counterpart functions will lead to the hypothetical evaluation conclusion (see Table 5.1) following from the first latent process: the foreign and the local actors are both capable of carrying out a task, but when they work together their capability is reduced. Going beyond engineering-related TA by insisting on the counterpart functions ignores earlier criticism of TA.

One factor for the insistence on counterparts is the neglect of questions about the appropriateness of TA and of the contributions of foreign experts. Without this negligence, the counterpart function would automatically become more complex. The cause of this relation is the on-going transition from the blueprint to a still unclear process mode of TA. This cause would also affect the endo- and exo-social processes, respectively, in the two cases. But another general orientation of the Handbook would affect them even more: the orientation to sector studies and country-specific institutional development strategies for TA. These strategies should contain, on the one hand, fragments of the use of technical content to manifest cultural distance (the exo-social process) and, on the other hand, fragments of the meaninglessness of technical content without origin (the endo-social process).

For example, the assessment of the IBRD's industrial technology support programmes (IBRD 1995) records an exceptional reliance on technology imports in Mexico despite the strength of technological research institutions.[64] This evaluation study could not offer a remedy from the evidence at hand. It could only confirm the failure to attain the 'traditional attitudes towards R&D'. In its new science infrastructure project, the IBRD has withdrawn from industrial technology financing as such. We conclude from this evaluation study that, while the endo-social process appears at the sector level, it could not be understood or reached at that level. Consequently, the Handbook's application to Autogeneración, which would integrate a sector-wide institutional component in its design, would not produce any changes to the latent processes as a result of the exchange dynamics.[65] The sought-after country-specific institutional development strategy for Mexico remains an open challenge.[66] Here, going beyond engineering-related TA through sector analysis indeed integrates earlier TA criticism. The appropriateness of foreign contributions required is enlarged beyond the individual project. But for management purposes, it is enlarged in the wrong direction.

There is no management guideline in the Handbook which would affect the third latent process – the interface between developers and developees. Together, they develop a system of references that permits them to establish pseudo-solutions to the contradictions in their relations. These temporary workable solutions allow them to function. This process could be accounted for in the Handbook, but with interfaces remaining in the professional memory of individual developers and developees there is no basis for deriving management guidelines.[67] The difficulty in defining terms of reference (TOR) for consultants reflects the interface dynamics. Using ever more refined TOR is an expression of the difficulty of accounting for the evolving contradictions in developer–developee relations. The Handbook suggests providing TOR for consultants (foreign and local) and for counterpart staff but does not indicate how these TOR differ.[68] The refining of TOR remains blind trial and error. The only direct expression of the third latent process in the Handbook is the suggestion that lengthy periods of separation between experts and counterparts for training purposes are to be avoided, in order to achieve the most benefit from their interaction (ibid.: 107). Suspending the temporary workable solutions in their relations appears to be disruptive. To advance the institutional development orientation of the Handbook, it could account for interfaces where experts act upon their relations explicitly and (if the opposition between the cases has wider applicability) interfaces where the experts cannot express why the other's judgement is inaccurate, because these characteristics are institution specific. The case studies show that these characteristics also appear in engineering-related TA. Institutional development objectives in TA add to the interface characteristics, but probably less so than commonly thought. In that regard, the Handbook still ignores the many calls for a code of conduct for foreign experts that have appeared in the past.

A final comment can clarify the difference between the Handbook's guidelines and the management goals suggested. The Handbook is based on the assumption that there are two schools of thought on understanding institutions. The first is the holistic one, where all institutions in a society evolve endogenously according to a cultural determination. The second is a rather behaviouristic one, where institutions can be singled out and changed deliberately based on superior understanding of the institution's rationale. Contrary to this view, the goals suggested imply that the development intervention itself, ever since colonial times, has become, and remains part of, the endogenous changes of all institutions of a society. Developers also exist in the native's world, and their contribution is partially predetermined locally before a foreign expert arrives. For this reason, the suggested management goals are closer to TA reality.

Summing up, the management goals identified translate into concrete management tools which are not used currently. Past reform efforts in TA management have so far resulted in changes in TA implementation, but there is no comprehensive analysis of the impact of these changes. Recent management guidelines (IBRD 1993) contain far-reaching changes that correspond to the latent processes but, ignoring the nature of these processes, these changes cannot affect them and instead re-enforce the latent processes.

7 Outlook

As in previous development decades, various forms of technical assistance will continue to appear in development policy and practice. Many powerful social, economic and environmental conditions in the 'global village' induce a sharing of technology. And as technology is – just like other social arenas such as language or politics – socially constructed, technical assistance will remain a vibrant field of human endeavour. However, technical change has been ineffective as an agent for social objectives. Ultimately, scientific courts and other democratic means will be established to harness technical changes. Nuclear and genetic engineering are already forcing social choices into the arena of technical change.[1] Technology-importing countries are in a potentially fortunate position[2] to incorporate social decisions also in lesser issues of technical change. Possibly, new institutions such as the Global Environment Facility (GEF) will also play a major role in fostering social choice in technical change.

Because technical change still resembles socially blind crazes[3] in the industrialized and developing countries, developers and developees struggle with the social meaning of technology. Furthermore, their technical skills contain values, meanings and biases which are variations of social meaning. Contending with the social meaning, or rather its absence, TA participants deepen contradictions between planning and practice and the gap between the insider and outsider perspectives. Practice always consists of the reproduction of social meaning and cultural distance that TA theorizing and planning cannot account for.[4] The causes of these contradictions are therefore as profound and potent as the reasons for sharing technology. While social science assessments can foresee the impact of technical change, the contradictions appearing in TA practice remain unforeseeable. Even with perfect planning, insider and outsider perspectives of TA differ sharply and contradictions arise. This point seems to be supported by the fact that, if the obstacles to effective TA were not deeper rooted than they appear, aid donors and recipients would not spend billions of dollars every year for inefficient projects.

The gap between the insider and outsider perspectives of TA events is also a condition for the latent processes because, without this gap, the cultural distance between local and foreign participants would not be problematic for them. Without the gap,[5] their exchanges, problems and achievements would lead to conclusions and lessons learned, either by the participants' own efforts or, if they too are implicated, by an outside advisor. For the latter, however, as non-participants, the outcome of TA is inconsistent and contradictory. Therefore, the determinants for this gap are the driving force for changes in TA practice. We have found three paradoxes in the outsider perspective: the first is between the confrontation on technology and the agreement over its accuracy; the second is between the accuracy and the irrelevance of the TA outcomes; and the third is between the participants' intention and their effects. The insider perspective contains the elements of these paradoxes. The principal limitation to evaluation and management of TA is the political and economic resistance to accepting and acknowledging this gap. Finally, the gap was also the historical condition for an observer to become a pawn for the other participants. As a temporary insider, I allowed them to denounce the contradictions between TA practice and planning to me and to add to their repertoire of influencing cultural distance (mostly to reduce it).

From this fieldwork, we have presented a picture of development practice and an interpretation of the events in the case studies. Development practice is one of the most complex fields of social reality, therefore it is necessary to present such an interpretation to practitioners, developees and developers, knowing that they work on their interpretations as much as we have done. The methodological choices, using participant observation, comparing diverse cases, global anthropology and the Theory of Communicative Action, are appropriate for developers and developees to reflect on their practice. These choices have led to results which should be accessible to development agencies. Developers' and developees' readings are an additional justification for the choices. The results are also a call for more applied research. Practitioner and research concerns are combined to generate new theory to help reduce this gap.

Efforts by anthropologists to control this gap or to translate between developers and developees appear to be futile.[6] Rather than assigning the equivalence of meaning between the local and the foreign perspective, translating implies shifting the labels 'foreign' and 'local'. Anthropology can elucidate fundamental conditions of society. For TA, this implies showing how cultural distance is encountered and reproduced by participants. The participants we have observed transcend TA (even if they were not aware of the planning of these projects). Thus, independently of the passion in official development rhetoric and discourses, TA practice has to be seen in its own right. Participants determine the extent of the analysis. The insider perspective animating practice is analysed in its depth. The 'development machine' (Ferguson 1990) organizing the developers was first described in the 1980s and was increasingly criticized during the 1990s. The pressure to reform, for example, the UNDP or the IBRD and the tacit knowledge within such development agencies is increasing. The social sciences can weaken the habits and open up the development machine. Research opportunities exist that enlarge participants' abilities to influence their cultural distance.

Our fieldwork approach started with an observer joining the implementation of a TA project and becoming a participant in the exchanges between developees and developers. We then concentrated on the uses of the observer by the other participants. This usage (or manipulation) of the observer position reveals the social meaning of developer–developee exchanges. We reconstructed these exchanges and found coherent results with respect to evaluation and management traits in development agencies. TA practice is thus a viable focus for anthropology. Without showing the cultural dimensions of the developmental objects, the encounter itself is a sufficient object for applied research. Neither the foreign nor the local knowledge[7] can explain the relationships among locals, foreigners and their know-how in a TA encounter. Taking part in the exchanges allows us to see how the participants construct their encounter, both in the subsistence context of Chad and in the industrialized context of Mexico. We did not assume that perfect communication would allow the power of the expert to be overcome. All participants influenced the power differential between foreign and local experts.

> Continual ethnography, we argue, is the key to successful, cost-effective project management, *because the social worlds within which development efforts take shape are essentially fluid.* ... Gone are the days when anthropology could be labelled 'long and lost'; gone too the time when policy-relevant research was invariably 'short and shelved'.
>
> (Pottier 1993: 7, 12)

We have shown what this fluidity implies for TA practice in an industrial context.

The question of whom an ethnographer speaks when describing TA events is less important when she/he speaks for her/himself in identifying the objects created in the encounter. These objects do not defend or question the developer or the developee, as they are not misunderstandings but reflections of latent processes. This opportunity for anthropological analysis had to be demonstrated. From this, it may be seen that essentially two opportunities are open:

- relate this fieldwork approach to institutional and organizational factors in development agencies;
- test different forms of restitution of the results to participants and the impact on their TA practice.

The wide-open insufficiencies of TA theory can then be filled. The latent processes identified can be taken up as concepts in progress in development agencies. They can become part of the participants' exchanges; the processes can be talked about. Although agencies cannot control these processes, their anticipation can replace the paradoxes in the outsider perspective which informs TA planning. The paradoxes remain, but they cause less damage. Rendering tacit knowledge expressible has been an important source of innovation in high-technology firms, introduced as 'knowledge management'. Correspondingly, reference to latent processes enables TA practitioners to innovate. Expressing their tacit knowledge concerning these processes changes the content and the organization of their work.

The expected changes in practice spawned by the latest IBRD Handbook on TA will actually reinforce the content process, the confrontation on technology and the agreement over its pertinence (p. 172). The intended changes in TA management are evident when the 1993 Handbook is compared with the preceding TA management recommendations (IBRD 1983).[8] References to individual behavioural attributes of foreign experts, to training in cross-cultural communication and to four TA models with corresponding delivery modes have disappeared.[9] The attention to institutional analysis, monitoring and NGOs has expanded. TA design criteria have remained almost unchanged, including terms of reference, training and counterparts. The shift away from engineering-related TA to institutional development is specifically cited in the 1993 publication as it is in the 1983 one. All in all, the goals shift, but the means remain. This will reinforce the content process as long as TA planning does not anticipate a bigger role for local and foreign participants in changing the objectives during TA implementation. Long-standing administrative practices in development agencies have to be challenged. 'While [since the New Directions period in USAID] changes can be made during implementation, they require written congressional notification' (Hoben 1989: 259), and the IBRD considers change to be a violation of the Loan Agreement (p. 171). From there, it is a long way to conceding that TA implementation itself produces changes in TA objectives in each case.

The content and interface processes are more affected by development agencies. They have less influence on the exchange process as it reflects social identity formation.[10] Experimenting actively with different management approaches, for example introducing participatory evaluation and planning methods, agencies can learn to influence the content process and the interface process. Some UN agencies and NGOs pursue new approaches. For example, UNDP's experience with national execution[11] shows how cumbersome this is. Previously implicit definitions of the roles of participants are open to negotiation and new procedures are necessary to do so. Even when TA management allows major changes, participants are slow to take advantage of them. We have seen that participants in Appui Technique could not allude to the interface but some Autogeneración participants did (John and Miguel). This difference affects which management procedures are feasible. National execution improves the individual learning progress of participants from local institutions and development agencies alike. However, improvements are gradual and the central question is whether an innovative agency will get to the stage where individual participants are enabled to share their learning, and the lessons learned from the implementation in one case can be separated from the attributes of a TA encounter and be used as input for another case.

The tools suggested do not concern the technical knowledge itself. For exo-social exchange processes, tools aim at shifting cultural distance away from the technical knowledge; and for endo-social exchange processes, tools aim to mark the context specificity of know-how. Similar categories can be defined for each TA encounter. When all participants recognize that these are means to improve the collaboration, they can work. They are effective if taken up by both sides, and if developers and developees are addressed together. The 1993 Handbook does not contain such tools and thus begs their consideration. Participants can therefore be encouraged to consider these tools, and development agencies should propose them when participants express their interest. This implies that development agencies should experiment with consultations between participants, for example through workshops and visits between unrelated participants, and with the encouragement of participants to call upon development anthropologists at their discretion. First and foremost, participants can observe their practice differently so that they can see the potential of such tools.

New policies, for example regarding local NGOs and participatory methods in development agencies, can reduce the gap[12] between the insider and the outsider perspectives. However, the encounter between developers and developees evolves on a much larger scale than the intended or unintended changes in TA practice. 'The postcolonial predicament is now constituted by an interconnected series of religious, political, economic and social dilemmas that are global in their scope' (Breckenridge and van der Veer 1993: 3). This predicament constitutes participants' struggles to overcome the dominant colonial Other. The developer protagonism and the developee identity affirmation remain the heart of their encounter. All participants we observed also lived this predicament by acting as one in front of the other group, unable to respond to individuals. They did not question their individuality; on the 'stage' of the project, the roles were predefined. A mutual, unwitting silence limited the construction of new social meaning, both between local and foreign participants and on each side of the interface. Changes in TA practice result from the combined influence of the post-colonial predicament and TA policies. The contradictions between theory and practice will remain and become even more profound when assertive individuals (such as Osama, Pascal, Ramón or John) try harder but fail nonetheless. The intensive practice of identity is the hallmark of today's world, and even more so in TA practice. In the long run, this will change TA more profoundly than changes in TA policies.

The outlook for reforms is not promising. The imperatives of TA are getting stronger,[13] public opinion is critical, budgets are decreasing and new operational ideas require experimentation. The uncertainty of TA effectiveness adds to the tense climate in many development agencies.

Operational changes abound but lack depth. We can stress three insights for a theory of practice: the fieldwork results and their reception, the evidence for the latent processes (both from cases and from IBRD management and evaluation documents), and new categories of management tools.

If all TA has occurred in a hit-or-miss manner, as various evaluations have shown since the 1970s,[14] then TA practice has escaped managerial efforts also because it is within a moving predicament. We have seen several reasons why this will remain so. Reforms of TA practice reflect the contradictions between planning and practice, but the reforms have been so unspecific that they resolve only the most blatant of them. Encounters between developers and developees reflect little of what is contained in planning, an inescapable fact. Reforms occur in a context shaped by the combination of independent and, most importantly, local political factors and these contradictions. For example, a participatory appraisal process can work well in one country and not at all in another. 'The divergence may have roots in the fact that the two countries differ in terms of the leadership styles they foster, with open dialogue being an intrinsic part of the Ugandan approach, while Tanzanian officials (extensionists, political leaders, researchers) prefer a top-down approach' (Pottier 1997: 219).[15] The country context, the type of organization, the economic sector and social groups all combine to determine the local context. Changes in TA planning might never attain a more effective mode than hit-or-miss. The question is: who is really in the drivers' seat when it comes to TA reforms?

The latent processes are defined by the participants' combined faculties. Only they can change the processes. Tools such as those suggested can only support them in this. The participants are in the drivers' seat and can be enabled to use the steering wheel. As Todorov has stated, all combinations of the three dimensions of cultural distance are possible, but some are more likely for particular encounters. Raising participants' awareness of the latent processes enlarges the combinations. This can only be supported by development agencies through the anticipation of the scope of such tools during implementation. A 'shopping list' of such tools can be provided with no obligation to use them. In so doing, the orientation of such tools is acknowledged. Otherwise, agencies can only radically modify TA planning to affect current TA practice.

Reducing contradictions between theory and practice requires feedback from practice. Feedback can improve planning but only indirectly improves the participants' ability to combine their know-how. The latent processes can be assessed by participants and there is no planning reason to ignore them. Potentially, participants can resolve the latent processes completely. It is less plausible that agencies actively address the gap between insider and outsider perspectives. We cannot establish empirically at present whether participants are getting better at achieving TA objectives. Current changes in the IBRD and many bilateral agencies indicate that individuals fail to grasp their experience. Therefore, it would be necessary to compare one participant's learning over several TA experiences and between participants. This is methodologically not feasible for us. In other words, we do not know how far the objectives are missed and how to count 'white elephants'.

We have called the embeddedness of the participants' efforts and reflections the idiosyncrasy of a TA encounter. Efforts to change TA practice logically start by addressing this idiosyncrasy. The more acceptance an agency develops for the idiosyncrasy of each TA encounter, the higher the likelihood that insights from one encounter become available for the next. By exploring the various manifestations of this idiosyncrasy from the individual level to the macrolevel, the structure of this book fosters this acceptance. Ultimately, the latent processes represent planning limits because planners have no access or influence on them. Even worse, a general definition of this limitation is also impossible. The idiosyncrasy of TA encounters is still the major challenge with which to confront both development agencies and development anthropology.

Development anthropologists have fostered some applied research by development agencies. However, the expansion of development anthropology during the 'New Directions' phase in USAID fell short of its potential. This was primarily because it was concentrated on the outcome of a limited cluster of rural development projects and less on planning and implementation. This presentation of fieldwork in TA has emphasized that anthropology can expand to other areas of development. Research on the outcome and the implementation of TA can be combined. The practice objects found there can become the limits to which development agencies restrain TA planning.

The other way to support agencies' acceptance of TA encounter idiosyncrasy is to underline dysfunctional evaluation and management traits. Here, we have seen that, even when defined from a small number of case studies, the latent processes reveal contradictions in evaluation and management. Sometimes, an agency renounces the developmental objective just as the IBRD has withdrawn from industrial technology financing in Mexico. However, in most cases, an agency maintains the objectives despite the ineffectiveness and continues with trial and error. Implementation tools should be readily adopted when they relate to TA outcome. On the contrary, improving evaluation is severely limited as the best evaluations confirm the difficulty of isolating an element of implementation from the individuals involved. Evaluations[16] will continue to be largely irrelevant for TA practice and function primarily as an operational tool within development agencies.

Development agencies are often averse to tools that imply participants should reorganize their collaboration because this resembles a step into the dark and implies political risks. The acceptance of idiosyncrasy is encouraged by the demonstration of new categories of tools, latent processes and extrapolations from them. There should be a demand for applied research, rather than a push from anthropologists. All sectors of development aid can be studied as practice. Given the depth of a TA encounter, applied research can offer objects of analysis which 'attract' agencies to the idiosyncrasy of each encounter. The ruralist image of development anthropology can certainly be overcome. The professional stakes of participants are another access problem in TA observation. The presence of an observer appears to involve risks for participants. We have seen that applied research can help participants to shape their encounter. Thus, larger projects than those examined here, involving higher risks and more individuals and organizations, can be studied with this type of fieldwork. When local and foreign participants know that an observer's presence has been useful to other participants, the access problem is reduced in the eyes of both the participants and their employers.

In both case studies and the participants approved the results from the observation and the response was positive (with the exception of Ramón). In their comments, the conflicts during their encounter took a more positive turn. They reattributed individual and institutional responsibilities for the outcome of the projects because they were able to grasp that structural causes had been important. The response from the development agencies was non-conclusive (section 3.3). Possibly, the hybridity of the results (containing developer and developee traits) limited the institutional response, but this was not logically consequential. A better reconstruction of the events (or with other methods) might lead to a different response. Leaving this possibility aside, one could infer from the difference between participant and agency responses that these agencies refuse to accept the fundamental status of implementation conditions, as if agencies would be ideologically opposed to these mainly local conditions and as if development agencies would insist on the superiority of, say, a Western, a capitalist, or a scientific perspective. This conclusion exaggerates the agencies' responsibility for the gap and is not valid for many. That the results are more useful to the participants and less pertinent to an agency also reflects that agencies cannot attain the latent processes. Because of the idiosyncrasy of each TA encounter, participants have little basis on which to compete with the arguments of planners. It is a fundamental necessity that agencies accord a higher status to implementation conditions and problems[17] than they have done in the past. Accounting for the gap and its many manifestations in all TA activities requires a considerable amount of what appears in agencies as blind faith, especially in local participants. Occasional references to 'local ownership' will not reduce the contradictions between planning and practice, and will serve mainly to immunize agencies from TA outcomes.

Discarding the paradoxes in the outsider perspective, reducing the dominance of TA planning (the project cycle) and then giving control over TA objectives to participants are difficult institutional learning advances that development agencies struggle with. Most of this simply reduces paperwork and technocratic mirages. Ceasing to burden participants is one stage, actively enabling them is the next. Operational changes in TA are always behind the individual learning by practitioners. Pascal, John, María and Mohammad were far more subtle with their know-how than the design of these projects anticipated. This is always the case. Participants are obviously in the drivers' seat as operators, but fundamentally also because of the social problematic of technical change. Participants' skills are always ahead of evaluation and management advances because the idiosyncrasy of each implementation is intrinsic to an exploitation of technology in a foreign society. Especially with industrial technologies, the incapacity of experts to separate the instrumental core from the social meaning of their knowledge creates the content process and thus TA practice becomes idiosyncratic. The analysis presented enables participants to invigorate planning, management and evaluation. The reconstruction of the cases enhances participants' learning. Synthesizing the conclusions for evaluation and management reform is a secondary task. First TA encounters change, or they do not, and then evaluation and management can encourage that. This remains the profound challenge for development agencies. First and foremost, participants must observe their practice in a different way to see the potential to change their encounter.

7.1 Research outlook

Changes in TA practices depend on the capacity of development agencies to come to terms with the idiosyncrasy of implementation and with their own institutional expressions of cultural distance despite the political conditions of North–South relations. Social science research on technology's contribution to development reveals the social conditions in which technical knowledge is created, circulated, adapted and applied. The latent processes we have defined show the obstacles to overcome and the limits of agencies' reforms. Therefore, the latent processes allow us to flag those social conditions of technical knowledge that are effectively attainable given past TA practices. For industrial technology in particular, these conditions are still largely unknown to development agencies. Because the idiosyncrasy of implementation was our prime target, our micro-observations are too empiricist to identify these conditions and avoid unrealistic conclusions. However, specifically designed comparative research can produce insights and advice in particular TA contexts. We highlight some striking reasons for doing so.

The macroeconomic focus of development policy has reduced the attention to technology since the 1980s. Nonetheless, the need to understand sectoral and microeconomic interventions is becoming more evident (Rodwin and Schön 1994). On that level, the contribution of technology to development is prominent. The unclear effectiveness of TA efforts has also contributed to the lack of research on TA. The methodological obstacles add to the generally difficult political economy in which TA research occurs. Outside development research, several schools of thought have been established whose insights are highly relevant to TA and whose methods and objects apply to TA practice. In general, these have not been utilized in developing countries only because their innovators work in the North. Science and Technology Studies, Learning Organizations and Organizational Development, Evolutionary Economics and Knowledge Society research will be applied to development assistance. The degrees of freedom of TA correspond in these schools to the 'seamless web between technology and culture', the 'symmetry between society and nature', 'deutero learning', 'fit between techno-economic paradigm and institutional setting' and 'materiality and sociality'. These schools of thought can magnify the social dimensions of technology development that agencies address.

Many of the earlier concerns regarding TA are reappearing under the headings of institutional development and globalization. The social dimensions of technology can only be understood in a particular context. TA implies the movement of technical knowledge from one context to another. Different forms of technical knowledge must be blended together through exchanges between local and foreign experts. The conditions for such blending can be defined (instead of a transfer of pre-existing knowledge) by observing such exchanges and linking the limits and opportunities of such exchanges to social processes. Two broad orientations for future research on TA practice are obvious: first, improving the demonstration of the idiosyncrasy of implementation to enable practitioners to reflect on their own work, and, second, advancing the definitions of the latent processes.

Both orientations hold considerable potential for theoretical and empirical extension. Practitioners relate case studies on TA implementation in those countries, sectors of the economy and specific technologies where they work more effectively to their own experiences. This represents the empirical extension, and the potential corresponds to the limits of the practitioners' agency, their social problems. However, practitioners can also consider theoretical extension and anthropology of development. In a given context, certain properties are thought of as intrinsic to a technology: who can possess technical knowledge, who creates it, how does it circulate in the global or regional market, and so on. Differences in these criteria between two contexts determine the tasks for TA practitioners. Theoretical extension also supports practitioners in speculating, innovating and experimenting.

Latent processes allow development agencies to understand the institutional dimension of TA. This implies habits and predispositions for project success and failure in development agencies and in institutions in developing countries. We have shown that the latent processes explain the dynamics of project implementation irrespective of the properties of technical knowledge in either of the two industrial contexts concerned. Each latent process can perhaps be distilled to a small number of possible configurations. These are likely to represent ideal-types, determined by larger historical changes in a society. The above-mentioned schools of thought can improve our understanding of the ideal-types of the latent processes. A TA effort in any sector, country or technology can include explicit reference to the latent processes likely to appear during its implementation. Practitioners always reinvent the historic context within their encounter and reinterpret their work accordingly. Development agencies might use the explicit reference in their relations with local institutions, with practitioners and with funders, depending on the respective sensitivity to social dimensions of technology.

The general director of the evaluation department of the IBRD sees the challenge to learn from development projects: '[development evaluation] is being increasingly pressed to consider the processes out of which development projects emerge and through which they are executed, the legal and regulatory environments in which projects operate, and especially the growing role that local authorities, non-governmental organizations, and other elements of the civil society play in development' (Picciotto in Rodwin and Schön 1994: 227). Technical knowledge is shaped by educational institutions, by companies and by professional and industrial associations, and the interactions of these are determined by explicit rules and, more importantly, by tacit rules, especially 'the traditions of co-operation, discipline and participation embedded in the local scene' (ibid.: 217). The same language is used by Klitgaard (Klitgaard in OED 1998: 73), an indication of the difficulties to be found in following up the analytical ambition. The IBRD is possibly well on its way to admitting the idiosyncrasy of TA and the autonomy of the latent processes, but the challenge of integrating sociological and anthropological insights into the economic logic of development agencies (and the project cycle) remains immense. Development practice and TA in industry, in particular, is a field where this integration can be pursued.

The present results are not specific to a development agency.[18] Having opened the empirical field, future research on TA practice can be more instrumental to particular TA policies and/or agencies. Such research is pertinent for all applications of technology in a new context in which the co-operation between different experts is necessary so that their accumulated know-how is also applied, whether a technology is used in a society other than the society of origin or whether it is used in a firm other than the firm of origin and irrespective of the scale of the technology. The exchange of know-how, in its complexity, between local and foreign participants contains all the necessary elements to account for the economic, social and cultural dimensions in the TA context. Based on participant observation, such comparative research is only feasible with on-going TA efforts and should constitute the experimental part of a TA policy or programme, allowing TA to be refined and to evolve by integrating those conditions that are beyond the reach of planning.[19]

The results invite the theoretical and empirical extension to economic and technological conditions. Different stages of import substitution and export policies correspond to differentials in technological capacities and the latent processes reflect these stages.[20] This allows technology-importing countries to articulate industrial policy for technological capabilities of different organizations, firms, universities, industrial associations and development assistance activities such as TA. The institutional side of TA has to be analysed jointly with the technological side. There are specific conditions, e.g. in energy, telecommunications, pharmaceuticals and other sectors. Comparative case studies similar to those presented here yield specific results, e.g. for a firm, a technology or a country context.

Particular opportunities exist in energy policy, for example. Climate change mitigation is a central arena where technology acquires new social meaning. Carbon dioxide emission rights have become a global concern, and the UN-FCCC[21] negotiations led to policy regimes and institutionalization. The sustainability paradigm contains highly productive contradictions between ethical principles of equity and resource management imperatives. Global ecological concerns increasingly challenge determinist theories of technology and, for example, 'leap-frogging' from coal to solar technology has to be achieved.[22] Sociocultural ends of technology found in energy TA can be taken up in the climate change arena, support policy development and, in return, become the objective of TA. For example, cogeneration, load management (in the power grid) or joint implementation in Mexico, India or Southern Africa have been hampered by the role of foreign technology that has been proposed. Autogeneración was limited by the Mexican experts' grasp of context conditions. TA could provide cogeneration technology for specific sectors by taking into account traits from PEMEX (Petróleos Mexicanos) and CFE,[23] especially gas and electricity supplies. This would not be done normally because it is technologically unnecessary. In addition, carbon dioxide emission reductions can be weighted for export earnings (Mexican oil) and opportunity costs of joint implementation investors. This would assist Mexican experts in defining context conditions and in demonstrating their contribution to 'national technological advancement' (similar to the competence shown by PEMEX and CFE) *vis-à-vis* Mexican firms. 'Mexican carbon dioxide' would be an element of technological capacity generation between Mexican industrial institutions and would render the latent content process in TA more effective to the TA objectives, especially because energy companies have played prominent historical roles in the industrialization in all countries.

Each field of industry posits economic and technological conditions for TA. Charles F. Sabel has shown that social learning makes a key contribution to developmental associations in general.[24] Japanese producer associations, Italian SME networks and German trade associations set up negotiation and co-ordination frameworks that allowed unprecedented technology co-operation to strengthen economic growth in these countries. The IBRD, UNDP and the Organization for Economic Co-operation and Development's (OECD) DAC currently advocate Sector Investment Programmes (SIP) to address systemic learning processes. Sabel demonstrated the relations between firms and the state necessary for the discursive formation of interests in such associations. 'Incalculably valuable non-pecuniary externalities' can be achieved by combining learning and monitoring in industrial relations through such associations.

> Individuals must interpret the general rules and expectations to bring
> them to bear on the actual situation. These reinterpretations proceed
> through argumentative encounters in which individuals attempt to
> establish an equilibrium between their views and social standards by

recasting both. It is this reflexive capacity to embrace different forms of self-expression that defines persons as individuals and creates new interpretative possibilities for society.

(Sabel in Rodwin and Schön 1994: 268)

Sabel applies concepts from general pragmatic theory to learning between individuals and between firms in a similar manner as we have done in section 5.4.[25]

Pragmatic analysis of social learning can explain why firms in some countries manage to pull themselves up by their 'bootstraps'[26] out of low-equilibrium traps through co-operation to improve production. While such learning cannot be predicted, it can be *diagnosed* once it takes place. The co-operation in TA rests, even more than 'bootstrapping' between firms, on emerging discursive means to translate know-how from the foreign to the local perspective. Language is so ambiguous in TA that meaning can be produced only co-operatively, through joint elaboration of a common framework of understanding in discursive conversation. Social learning in TA reflects technological capacity in an economic sector. TA can be adjusted to the state of affairs in developmental associations in industry. For example, 'bootstrapping' lessons learned in Japanese industry can be adapted to the technological learning in several developing countries (Kaplinsky 1994). Understanding the sociocultural specificity of TA outcomes enhances the functions of local developmental associations identified by Sabel. Defining the roles of experts, their mutual responsibilities, affiliations, information channels and products addresses the learning and monitoring functions of the industrial organizations involved.

These opportunities for research in energy and industrial policy illustrate how on-going TA efforts can be assessed and, with an expanded empirical basis, how they can enable TA effectiveness to be improved in specific industrial conditions. TA implementation ceases to be a 'black box', and the latent processes are actively shaped by TA practitioners. Our results can be connected to policy debate and development. TA outcomes become more significant for industrial development than for narrowly defined microeconomic TA objectives. Although these microeconomic objectives are more reliable in industry than in agriculture (especially small-scale farming), and thus the calls for impact assessments and better evaluation have not appeared that much, the confusion, blindness and inefficiencies from the unknown role of knowledge and expertise in TA for industry are just as open to enquiry. TA practice is a major gateway to attaining the potential of technical knowledge in development.

The proposed TA ethnography explains, for example, that the content process is created by claims that sociocultural ends should be part of the instrumental core of technical knowledge. These habits lead to an overdetermination of the parameters of a particular technology, often affecting the outcome negatively. The content process changes when a TA project is moved from a governmental institution to an NGO, from a bilateral to a UN programme, or from a university to a private firm, for example. Such organizational modifications alter the discursive formations available to practitioners and therefore alter the three latent processes. Understanding how these processes change improves the accuracy and adaptation of such modifications. Context-specific management tools can be derived for each latent process configuration (content, exchange and interface) and form of cultural distance (alterity). Shifting cultural distance away from technical knowledge and separating the value axis from the relational axis corresponds to the exo-social exchanges in Appui Technique. The same management tools are useful for other exo-social exchange encounters, but other aspects of cultural distance can result in a different set of management tools. Defining tools in the abstract for each axis would be misleading, however our fieldwork approach allows us to avoid positivist exaggerations. Beyond a specific TA encounter, the first constant for management tools is any combination between development agency/economic sector in a country. For example, UNDP-led efforts in banking in Thailand would form a pair where a specific subset of management tools can be defined as a result of observing implementation. These are not necessarily adequate for another agency or another economic sector.

The monitoring and reorientation of on-going TA projects are the most promising research opportunities. The methodology proposed allows us to harness the dynamics of the implementation of a particular TA project, identify appropriate organizational and technological modifications, and improve it with respect to institutional development and capacity-building objectives. The most obvious policy elements and research objectives are organizational learning, technological capacity building, incentives and performance indicators for practitioners, research and development design, ownership of TA, typologies of management tools for specific industries, sector investment programmes, international joint ventures, adaptation of technology, the role of local knowledge and forms of local expertise. The reorientation of TA implementation can encompass the division of responsibilities and roles among practitioners, consultation and communication design, task structuring, budgeting, information systems design, administration and reporting routines. The results can be used to reorganize the TA project observed and inform similar TA efforts.

Project preparation and evaluation, i.e. before and after project implementation, are secondary research opportunities. Frequently, TA design combines the direct impact with structural and dynamic impacts. The latter enables local practitioners to learn, apply the know-how to other contexts and raise local professional standards. This dynamic impact is the core objective of TA. The two case studies highlight that the dynamic impact is not connected to present TA designs. The monitoring of on-going TA projects offers the possibility of attaining the dynamic impact. It is additive to classic impact and cost–benefit studies, notably because it allows us to separate structural from individual causes of TA outcome.

Since the practitioners are in the drivers' seat, applied research can only follow and bring out the changes in TA practice. The extent to which applied research on TA is called upon depends on the commitment of development agencies to work on the institutional limits which constrain their operations by reducing the parameters of TA they acknowledge. Those that they assume they have to exclude continue to be labelled local, cultural or psychological and serve to justify development agencies' existence while, at the same time, they perpetuate their ineffectiveness. We have demonstrated with renewed vigour that the relations between local and foreign practitioners are all-inclusive and everything relevant to TA is attainable through them.

Appendix 1

Presentation of Appui Technique and this research in a journal sponsored by the French 'Ministère de la Coopération

Les formateurs à l'école de l'apprentissage sur le tas

Les artisans en menuiserie métallique sont désormais si nombreux à N'Djaména qu'ils ont de plus en plus de difficultés à trouver des débouchés. Pour aider les entrepreneurs à diversifier leurs produits, une Cellule d'appui technique aux micro-entreprises a été créée. Les actions de formation menées par la CAT empruntent largement aux 'méthodes' d'apprentissage sur le tas. Thomas Grammig, un chercheur allemand qui étudie l'apprentissage traditionnel et les transferts de technologie, a suivie ces actions. Ses premières réflexions complètent ici le descriptif que les responsables de la CAT ont bien voulu nous faire parvenir.

Ils n'étaient que trente à N'Djaména il y a dix ans, ils sont maintenant deux cents. Tous à proposer les mêmes services de menuiserie métallique. Bien sûr, la multiplication de ces ateliers artisanaux prouve que la demande est forte, que le marché existe. Mais à n'offrir que des produits semblables, de qualité médiocre, les artisans oublient de fidéliser leur clientèle. Et risquent chaque jour la faillite… En attendant, dans les ateliers, on vivote au rythme des commandes vaillamment décrochées, comme c'est le cas chez monsieur Abdou. Situé sur un axe principal menant au centre de N'Djaména, son atelier n'a dû son salut qu'à une commande inopinée, alors que la clé était déjà sous la porte!

Dans ce contexte, la Cellule d'appui technique aux micro-entreprises a pour but d'aider les artisans à diversifier leur production. L'action se mène sur plusieurs fronts. De nouveaux produits sont d'abord mis au point qui tiennent compte de deux élément: le marché local (une réelle demande) et les conditions habituelles de production des ateliers (savoir-faire et équipement). Pour faciliter ensuite le transfert de technologie, qui permettra aux artisans de fabriquer eux-mêmes ces nouveaux produits, la CAT met en oeuvre des actions de formation-action spécifiques inspirées du monde d'apprentissage traditionnel. Un troisième volet aborde les questions de l'organisation professionnelle et de la promotion commerciale.

L'originalité (et leur intérêt) des actions menées par la CAT est que celles-ci reposent sur une grande connaissance du milieu artisan. L'un des responsable explique: "Les micro-entreprises fonctionnent selon une stratégie permanente de survie économique et de recherche de liquidités qui permettent à l'atelier de travailler. La préoccupation de l'artisan du secteur informel est de vivre *au quotidien*, avec les qualifications techniques qu'il a le plus souvent acquises sur le tas et qui suffisent à la clientèle habituelle, plus soucieuse du moindre prix que de la qualité." La mobilisation des artisans et leur participation aux actions de la CAT seront donc liées en priorité à l'intérêt économique à court et moyen termes qu'elles peuvent représenter.

La micro-entreprise, un univers à comprendre

Une autre caractéristique importante à comprendre pour agir dans le milieu de la petite entreprise est la perception qu'a l'artisan de son travail et de son statut. Les ouvriers de l'atelier de monsieur Abdou, par exemple, reconnaissent avoir choisi le métier de soudeur pour être indépendants et travailler à leur compte. Pour eux, l'idéal du patron est aussi attrayant que le revenu potentiel. Le côté créateur du métier est également un élément important de la motivation, même si le peu de moyens limite les activités novatrices. Ni le produit fabriqué final ni la machine utilisée ne sont porteurs d'une qualité intrinsèque. Ainsi les ouvriers de l'atelier ont l'habitude de travailler avec un poste à souder qui, n'étant pas reglable, valorise leur dextérité dans le contrôle manuel de l'électrode. Ce que le client achète, en premier lieu, c'est cette dextérité de l'ouvrier. Et chacun d'affirmer: le client doit bien connaître les ouvriers qui ont fabriqué ce qu'il achète.

Cette relation entre l'artisan et le client prolonge, en quelque sorte, les rapports personnels qui priment au sein même de l'atelier. Le patron est plutôt le meneur d'une équipe plutôt que le chef d'entreprise. Et c'est le fait de se percevoir comme une ensemble organique qui permet à l'équipe de fonctionner. L'organisation du travail découle de cette conscience commune. Dans l'atelier, les outils comme les produits semi-finis sont à la disposition de tous. Chaque ouvrier entreprend le travail qu'il perçoit comme nécessaire. Toute la production se déroule sans autre instruction du patron que celle touchant à définir le produit final, et l'aboutissement du travail résulte d'une volonté d'agir ensemble. Le rôle principal du patron est de trouver des commandes, pour donner au travail à tous.

Apprendre, c'est observer

Le mode d'apprentissage est le reflet de ce mode de fonctionnement. Dans un garage, selon monsieur Abdou, "un apprenti qui ne sait pas conduire attendra une opportunité, la clé oubliée dans une voiture par exemple, pour essayer de la faire démarrer comme il l'a vu faire par les autres. Le patron ne lui expliquera rien." D'une certain façon, l'apprenti doit voler un savoir que le patron n'est pas enclin à lui donner trop vite …

Dans le secteur informel, le niveau scolaire ne dépasse que très rarement l'école primaire. La plupart des ouvriers et des patrons ont appris leur métier en travaillant dans les ateliers. Cela implique que leur savoir-faire se limite souvent à un produit spécifique et à l'acquisition des gestes nécessaires. La durée de l'apprentissage est estimée à environ cinq ans. Dans certains ateliers, où quelques ouvriers ont pu bénéficier d'une formation formelle, il est intéressant de remarquer que la durée de 'formation' d'un apprenti est réduite à deux ou trois ans.

La formation de la CAT tient compte de tous ces éléments. Les stagiaires travaillent eux-mêmes, avec leurs outils habituels, sur les modèles à produire ultérieurement. Les instructions ne sont données qu'au fur et à mesure de l'avancement de la production. La réalisation des prototypes repose sur un consensus qui s'établit entre les stagiaires et les formateurs au fur et à mesure de la fabrication. L'explication abstraite d'un travail à faire est une chose que l'on évite à la CAT.

A la fin du stage, les réactions des participants sont unanimement positives. Un facteur pour cette unanimité est peut-être la ressemblance entre la pédagogie du projet et le mode d'apprentissage par lequel tous les participants ont été formés. Ils estiment tous qu'il est indispensable d'assumer toute la fabrication du début à la mise en marche du produit final. Certains apprécient cependant que des notions plus abstraites puissent être introduites au fur et à mesure de la production. Cette appréciation d'ensemble reflète leur souhait de se baser sur le mode d'apprentissage qu'ils connaissent mais en même temps, d'aller au-delà de leur expérience. C'est ainsi que l'organisation de ces transferts permet d'avancer petit à petit vers l'utilisation d'outils plus complexes comme les plans et les croquis.

La connexion entre le travail en stage et le travail habituel dans l'atelier n'est pas toujours ressentie comme évidente. Néanmoins quelques uns disent comprendre mieux, grâce au stage, certains aspects de leurs propres fabrications. On peut donc espérer que, à terme, leur produits habituels pourront également s'améliorer.

C'est dans ce contexte que la CAT a aidé à la fabrication d'un certain nombre de produits: moulin à pâte d'arachide, broyeurs de céréales, matériel de bureau, etc. (voir illustrations), autant de réalisations auxquelles ont participé plusieurs artisans-chefs d'entreprises à N'Djaména. Et si elles sont encore limitées en nombre, elles offrent déjà des avantages certains.

Mais sans suivi technique, les fabrications récentes en atelier risquent très vite de sommeiller dans un coin, voire de péricliter rapidement. Aussi, la CAT a mis en place des structures d'accompagnement chargées de veiller à la bonne continuation des projects. Il n'est pas question pour autant de prendre en charge les responsabilités des artisans-patrons. Simplement, si un problème surgit, elles seront là pour les aider à trouver des solutions durables, que les artisans pourront s'approprier vite. Cela va de l'aide à l'organisation des micro-entreprises en groupement d'intérêt professionnel pour répondre aux appels d'offre, aux facilités d'approvisionnement au moindre coût, en passant par la promotion des productions artisanales, l'organisation des sessions de formation, etc.

C'est souvent qu'un artisan expose atelier le prototype d'une nouvelle fabrication. S'il n'a pas rapidement de commandes, il le découpe pour réutiliser la matière première. Espérons que ce n'est pas le cas de cette micro-enterprise togolaise.

Les petits ateliers du secteur informel sont loin d'être aussi imperméables à l'innovation qu'on veut bien le dire. Et la participation de monsieur Abdou le prouve largement, lui qui aurait pu ne jamais remettre en cause sa situation d'homme établi, aussi précaire fut-elle. Une aide correctement orientée, qui prend en compte les modes de vie et de fonctionnement des ateliers, peut leur permettre d'évoluer vers une forme d'entreprise plus élaborée. Il s'agit avant tout d'être plus partenaire que donneur de leçons.

Point de vue du réseaux

CONTRIBUER AU DEVELOPPEMENT DES MICRO-ENTREPRISES
L'apprentissage est le mode de formation le plus pratiqué en Afrique; parfaitement adapté à la pratique culturelle, sociale et économique des artisanats, il en reflète autant le dynamisme que les limites, en terme d'amélioration de la qualité des produits et des services.

A l'heure des PAS (politiques d'ajustement structurel), le secteur de l'artisanat voit enfin reconnaître son énorme contribution au produit national, et les politiques publiques de formation, qui commencent à s'y intéresser, envisagant une reconversion partielle des centres de formation technique et professionelle vers l'apprentissage. Mais attention! Les apprentis et leurs maîtres-artisans n'ont rien à voir avec le public scolaire. Avant de vouloir les aider, il faut reconnaître leurs qualité, comprendre leurs motivations, entendre leurs demandes.

Certains projets très localisés d'appui à de petits groupes d'artisans méritent d'être étudiés parce qu'ils peuvent fournir quelques premiers éléments sur les pédagogies tout à fait originales qu'il faudra mettre en oeuvre.

First published in
Enseignement Technique et Développement Industriel,
bulletin no. 4, October 1992, pp. 24–26.

Appendix 2

Nine-point declaration drawn up by the Chadian artisans

1 Nous sommes convenus de constituer un groupe d'artisans-soudeurs, nommé un représentant de notre groupe, le groupe a choisi M. Osama, comme responsable du groupe qui doit faire la liaison entre le groupe et le projet.

2 Le groupe des artisans demande au projet, s'il y a une commande quelconque que cette demande ne soit attribuée à aucun artisan qui n'a pas suivi la formation.

3 Le groupe demande l'appui du projet pour faire une publicité à la radio ou à la télévision de tout ce qu'on vient de fabriquer; et le groupe préfère que la publicité soit au nom du groupe des artisans.

4 Après la publicité, le groupe propose fabriquer tout les trois prototypes avec l'aide du projet, après louer une boutique pour les exposer et les vendre.

5 Le groupe demande au projet les plans de tous les prototypes, et ne pas les remettre à aucun des autres artisans qui n'ont pas participé à la formation.

6 Le groupe des artisans demande au projet d'adresser aux bailleurs de fonds de trouver des commandes de charrettes et autres.

7 Le groupe demande au projet, s'il y a autres prototypes à faire qu'il est prêt pour les apprendre.

8 Le groupe demande au projet de faire un cours d'apprentissage de lecture des plans.

9 Le groupe demande au projet que M. Dambai soit conseiller du groupe.

Comment

This declaration was read out loud to the experts in the presence of all the artisans. It was never followed up by the artisans nor by the experts as it reflected mainly Osama's perspective. Osama had invited all the artisans to his house to draw it up but the artisans were unable to agree on a common position on how to relate to the French experts. This was implicitly evident to both the artisans and the experts. Osama's influence was also the result of his capacity to write and the errors in the text indicate his French writing skill.

'Le groupe' and 'le project' were accepted as entities by the artisans and the experts. The artisans attempted to define the relations between the two mainly by demanding those services from the project that they needed for their commercial success. While demanding services, the group preferred the advertisement to be made in their name. They assumed that the experts knew better how to advertise. Even Osama was not sure how the group should relate to potential buyers, especially foreign donors (*bailleurs*).

Notes

1 Introduction

1 An individual project is the basic building block used currently for most types of policies, institutions and technologies.

2 According to dematerialization studies, technology's role in growth is further increasing. For a discussion of the data, see Inkster (1991) or Landes (1998).

3 Outside development agencies, technology transfer is alternatively criticized by some for ultimately being a vehicle for recipient country dependency on the North, whereas others assert a potential to emancipate the South, depending usually on the political bias of the critic. Technology transfer agonizes by being a convenient adversary.

4 The authors used data from different years in the 1980s because the published accounts vary considerably. Time series are only given for OECD countries, from where the total number of technical assistance personnel has slightly increased between 1960 and 1990. Assuming a developer–developee ratio of 1:5, technical assistance employs 2 million individuals. Different categorizations are used for expenditure by donor agencies; a total $US50 billion annually is often cited as estimated total volume of technical assistance.

5 Different versions of the constructivist approach are pursued by researchers such as Trevor Pinch, Thomas Hughes and Wiebe Bijker (Pinch *et al.* 1987) and Bruno Latour (1987). While they do not form one school of thought, these approaches have gained acceptance over older materialist and culturalist approaches. All of them hold that technological change is not an external process, alienating human beings and subjecting them to exploitative interests, but an intrinsic part of human endeavours, reflecting general social and political processes.

6 The 'New United Motors Manufacturing Inc.' (NUMMI) in Fremont, CA, began production in 1984; see Dennehy and Cheng (1996), Wilms *et al.* (1994), Adler and Cole (1993) and Kraar (1989).

7 See, for example, Hamnett (1970), Morrill (1972), Harris (1977), Spitzberg (1978) and Hawes (1979). Fry and Thurber (1989) revived the approach, but their parameters such as dedication, commitment and Protean adaptability appear again too composite to be useful.

8 More information on the Social Development Action/PADS project has been made available on http://www4.worldbank.org/sprojects/Project.asp?pid=P000520.

9 Susan George and Fabrizio Sabelli provided the best account of the nature of this institution (George and Sabelli 1994). The fact that both projects discussed in this study were funded by the IBRD is incidental, and they have been chosen simply because of their accessibility. These projects were accessible because of prior contacts from my engineering career.

10 Smaller parts of that loan came from the European Union (EU) and the French

coopération. Earlier experiences in which the introduction of SAPs destabilized governments and even caused food riots certainly played a role.

11 More information about the Hydroelectric Development Project has been made available on http://www4.worldbank.org/sprojects/Project.asp?pid=P007609.

12 Today's interest in cogeneration (or combined heat and power, CHP) is also created by the need to reduce CO_2 emissions in order to limit global climate change.

13 'Actor' as a general label corresponds to the basic orientation of this study. It refers to individuals. Developer, developee and expert are used as labels only to allude to the behaviour of an actor. Similarly, 'event' is used to refer to 'what happened'. Instead of writing about 'the project', a semantic obstacle, writing about 'events' avoids giving 'what happened' a singular frame. Details about the actor concept are presented on pp. 90–92.

14 Projects perpetuate more projects. Projects are the extension of development discourse which circulates among foreign experts. Because they are unable to understand the local reality, projects have no other impact than unintended consequences.

15 The combination of power and cultural distance leads to painful errors. The Spanish Queen Isabella la Catholica sent the first Indians, brought to her by Columbus, to Rome, asking the Pope for advice whether these were humans or animals. A vice-president of the IBRD, Lawrence H. Summers, stated 500 years later that it would be economically efficient to dispose of the solid waste from the industrialized countries in the developing countries. The office memorandum was leaked to the press and then widely discussed, see, for example, Pearce (1992).

16 See the evaluation report from the Nordic agencies cited on p. 121.

17 If the feasibility studies produced in Mexico had led to investments, PEMEX (Petróleos Mexicanos) would be exporting more oil, the stability of the electricity grid would be higher, the energy costs in Mexican industry would be lower, the experts would have more contracts and the standing of the Mexican Energy Agency would be higher. In Chad, the domestic production of agricultural machines would have reduced the costs to the Chadian farmers, freed scarce foreign exchange for other goods, increased the profits of the artisans (in the informal sector) and expanded employment in N'Djaména.

18 He explained this when I visited his workshop one evening.

19 Note that the different reasons for the dynamics of implementation are interrelated and not fully separable analytically. Describing the effects of implementation will not help to disentangle them.

20 For example, we will see how an interface appeared during the events between local and foreign actors and how this interface then solidified these limits. An interface can also be described as an interactive filter. Graphical schemata for the interface in each case are explained on pp. 64 and 101. But more than this, the differences between insider (actor) and outsider (planner) perspectives are reinforced by development agencies' operational rules. For this reason, evaluation and management must be considered in this study.

21 He borrowed the term from the pyramid of Cholula, described in Octavio Paz' work on premodern Mexico (Berger 1974).

22 Realistically, writing a management manual is as reductive as analysing technology transfer in terms of payments for patents. Manuals are of little use because the actors cannot disentangle the message from the medium.

23 One touched an ear, the other a horn and the third a leg. Speaking about their impressions, they wondered whether they had touched the same animal because their impressions were so different.

24 The index allows the reader to locate these for each actor, but reviewing all appearances of an actor does not reconstitute his or her personality. A more

detailed interpretation of the project results in a case study that is only interesting to foreign experts. Being at the root of the events, only they need a singular 'correct' account of technical assistance.

2 Development anthropology

1 For example, Apthorpe and Gasper (1996), Bossuyt *et al.* (1995), Gow (1997) and Grillo and Stirrat (1997).

2 See Gardner and Lewis (1996: 26–49) for an historical account and the running concerns of applications of anthropology in colonies and in development practice. Unfortunately though, much of the social science produced in colonies and in aid programmes remains inaccessible to scholarly work.

3 Funding from the Rockefeller Foundation and the Ford Foundation has also played a role, in particular via the Consultative Group on International Agricultural Research (CGIAR) research centres (Rhoades in Green 1986: 26).

4 See, for instance, Green (1986), Wulff (1987), Bernard and Pelto (1987), Little and Horowitz (1987), Brokensha and Little (1988), van Willigen (1989, 1991), Chaiken and Fleuret (1990) and Salem-Murdock and Horowitz (1990). These are only some of the more recent publications from these authors. Much is published by Westview Press.

5 'In the early period of AID's assistance to Africa, from 1951 to 1972, US assistance could be characterized as a "bottleneck" strategy. Under this approach, scarce US development assistance was targeted on efforts to remove specific impediments to economic growth. Technical and capital aid were directed at physical and human infrastructure development, including agriculture and education, but also included larger-scale capital investments in transportation, such as roads and dams' (Greeley in Green 1986: 232). For a detailed analysis of the institutional situation of AID, see Hoben in Berg and Gordon (1989: 253–278). He sees 'New Directions' as being more influenced by the Vietnam War and the political climate in Congress; neither the appearance nor the disappearance of it was a genuine turn of development thinking or interest.

6 In applied anthropology, one finds similar biases to development anthropology. If an equally comprehensive source book was compiled in development anthropology, the biases would appear even more marked.

7 Consider the following, for instance: 'I was once told that a year-long position for an anthropologist would be written into a project extension if I would take the job; otherwise the position would be for six months or less and somehow combined with a technical position' (Koenig in Bennett and Bowen 1988: 359). I assume many more development anthropologists have had such experiences.

8 Starting from the perspective of the dominated and moving upward, looking at the sources of power. Koenig (in Bennett and Bowen 1988: 345) relates the call for studying up to Laura Nader (1974) 'Up the anthropologist – perspectives gained from studying up', in Hymes (1974). The profession is struggling with the new policies in the US development agencies.

9 'Oh, I've heard the stories at every meeting so far this year: anthropologists concealing their degrees in order to find jobs. Struggling to get secondary training or cast ourselves as nutritionists, economists or ecologists – anything to prove that we're not *just anthropologists*' (Keisman 1997).

10 Those with academic credentials tended rather to return to more academic work.

11 It is important to recall the basic questions, my technical credibility (i.e. engineering degree) was necessary in order to gain access to the two projects studied. Without that credibility, the presence of a social scientist was institutionally impossible. Once I had gained a role within implementation (section 3.2), the credibility was irrelevant, the access condition had disappeared. Upon leaving Appui Technique, nobody reclaimed technical results from me.

12 See Schönhuth (1991) for the case of the German governmental aid agency GTZ.
13 This 'natural' outcome rather results from 'social scientists being latecomers to agricultural research programs. All but one of the International Agricultural Research Centres and most national research institutions now employ economists, but few have anthropologists or rural sociologists on their staff. Few social scientists have worked directly with biological scientists in the development of new technologies' (Gow 1991: 4).
14 Mantra, or development liturgy, is a stronger metaphor than buzzword or fad, it implies a distortion of perceptions.
15 See Coreil in van Willigen (1989: 143–157), Warren in van Willigen (1989: 159–178), Chaiken in Brokensha and Little (1988: 237–249), Davidson in Wulff (1987: 262–272), Green (1986: 107–120) and Enge and Harrison in Green (1986: 211–222).
16 Assuming a qualitative difference such that a hierarchy between different social spheres becomes impossible.
17 Every case study in this work ends with a description of the 'anthropological difference'.
18 The case studies for the volume edited by van Willigen (1989) were elaborated with the focus on the concept of knowledge utilization. Nonetheless, the 'success of policy implementation depends, to some extent, on a clear commitment to the common good as well as on an understanding of the political processes within and among representatives of these organizations' (van Willigen 1989: 313). The attention to the research process cannot replace the normative orientation which remains a central element in the process.
19 Translations and insertions (in square brackets) are mine.
20 'Eminemment passionné est le débat – ou l'absence de débat – entre ethnologues et développeurs. ... De façon générale, au sein de la corporation des anthropologues, il n'est pas de bon ton de se mêler au monde des développeurs' (Amselle 1991: 17–18). On the fiftieth anniversary of ORSTOM, 'Sciences Hors d'Occident au XXie siècle', a round-table discussion on 'The difficult dialogue between researchers and development actors' concluded that, in order to start a fruitful dialogue, it was first necessary to stop feeling guilty (ORSTOM 1996: 115–128).
21 French development agencies rely first of all on ORSTOM (recently renamed Institut de Recherche pour le Développement; IRD), the national research agency for the former colonies. Georges Balandier, head of social science in ORSTOM, head of the Africa department at the Ecole des Hautes Etudes en Sciences Sociales (EHESS) and professor at the Sorbonne, would probably have prohibited work for development agencies, as was done in the USA during 'New Directions', see also Gleizes (1985).
22 These national factors are attenuated with the increasing importance of multilateral donors such as the EU and the IBRD.
23 Possibly, it would have been better to continuously review the ethical principles that the Society for Applied Anthropology has given itself (the main forum for development anthropology), instead of defining them in 1983. Furthermore, they ought to be articulated to particular contexts of development practice and separated from the choice of the object. For a comparison, Bulletin no. 20 of the Association Française des Anthropologues (1985) was entitled 'Recherche et/ou Développement', the question of whether there was a contradiction between the two was left open.
24 On the individual level, an anthropologist not sharing the developmental perspective of a donor sees his products ignored. On that level, there is a connection, but on a macrolevel the myriad of potential contributions from anthropologists never appear. These potential contributions move out of sight

from an institutional perspective, the donor perspective and the scientific discipline's perspective.

25 Key publications which keep the discipline focused on itself are Sperber (1982), Fabian (1983) and post-modern anthropology in the USA.

26 The major publications of these anthropologists are Ferguson (1990), Long and Long (1992), Hobart (1993), Pottier (1993), Sabelli (1993) and Poncelet (1994). Most of these authors have been working for development agencies for certain periods and have then retreated to academia to elaborate on more pertinent analysis of their experience. Their closer associations, for example the EIDOS network, created in 1985, were formed for reasons of publication. Their scientific work and progress have in most cases been achieved individually.

27 In 1972, Marc Augé, using the example of Samir Amin, illustrated what happens when anthropology naively imports the objects of development economics. Ten years later and after the Socialist election victory with François Mitterand (1981), the *Revue Tiers Monde*, XXIII/90, contained an outline of a sociology of development, then explicitly judged feasible by Pierre Achard. It was hoped that French foreign policy would accommodate research on development practice. There is no mention of US development anthropology. Before Augé, Gérard Althabe had already reconstructed an anthropological account of developmental failure in the post-colonial conjuncture (Althabe 1968). Augé and Althabe both argued that the contradictions of development practice are insurmountable: 'l'observateur devient l'acteur d'un processus qui lui met au centre de l'univers villageous. Il devient un ennemi à abattre pour les développeurs' (Althabe 1968, 1969: 313).

28 His theoretical references to the sociology of Giddens, Schütz and Luckman are more important than those to anthropologists such as Gluckman. In Long and Long (1992), he refers to a Wageningen connection of all contributors.

29 'It is important to stress, however, that an actor-oriented approach is not action research, but rather a theoretical and methodological approach to the understanding of social processes. Nevertheless, we believe it has implications for development practice, in that it has a sensitizing role to play *vis-à-vis* researchers and implementers – both social actors in their own right' (Long and Long 1992: 271).

30 An important trap to be avoided. As Sabelli pointed out, a farmer's knowledge cannot be defined in relation to developer–developed interactions in the same way that a farmer's knowledge is described when understanding the reproduction of rural societies (Sabelli 1993: 34). His assessment of the French publications uses Georges Balandier as the main reference, whose problematic of the *situation coloniale* continues to guide French anthropology.

Sabelli cites Claude Javeau 'une science n'est pas seulement une épistemé, mais est aussi une tekhné ... ce couple est indissoluble au principe de toute construction scientifique' in his methodological reflections (Sabelli 1993: 97). Whereas Richards refers to Stephen Marglin, 'the historic localized peculiarities that led to a strict segregation between episteme and techne as forms of knowledge in Western society' (Richards in Hobart 1993: 61). The precautions Sabelli takes against his own scientific position are positioned by Richards between Western society and non-Western society. The scientific context of their efforts is different.

31 Long explains that actor-oriented approaches have always been a counterpoint to structural analysis in the 1960s and 1970s, 'but they fell short because of a tendency to adopt a voluntaristic view of decision-making and transactional strategies ... and some studies foundered by adopting an extreme form of methodological individualism' (Long and Long 1992: 21).

32 Klitgaard refers to African writers such as Ali Mazrui, Claude Ake, Henry Bourgoin, Daniel Etounga-Manguellé and Axelle Kabou. It seems that cultural management arrived in development practice 15 years after it took off in industrial management and corporate culture research.

33 Totalization is an analytical perspective, in which each culture is envisaged as a homogeneous and unitary ensemble. The extreme form– 'The' Indian of such and such island does ... – is a classic rhetorical figure whereby the writer claims control of the object of the analysis.

34 Sabelli cites Switzerland, the Netherlands and Sweden, where the stress on participatory development created interest in social science methodology (Sabelli 1993: 76). See also Schönhuth (1991).

35 What Michael Horowitz has termed 'heartland anthropological issues as the conditions for transhumance, and floodplain production systems' (Horowitz 1994: 9). Such claims identify effectively the kind of anthropology that Horowitz is pursuing.

36 For example, only anthropology can provide a key parameter in the Brundtland Report *Our Common Future*, the definition of future generations' resource needs. And only anthropology will provide the tools to speak about biodiversity and about the kind of nature remaining after changes in the earth's climate.

37 Chauveau in Boiral *et al.* (1985: 163).

38 According to Allan Hoben (in Berg and Gordon 1989: 271), USAID was less successful in pastoral livestock development, crop production, integrated rural development, seed multiplication and agricultural research (where anthropology could have had more impact) than it had been in higher agricultural education and rural infrastructure (see also Bennett and Bowen 1988: 12).

39 'In one form or another, large-scale engineering projects are very difficult to stop and have very little, if anything, to do with structural reforms in political economies at national and international levels. Let me elaborate. When the World Bank decided not to fund the Kossou Dam in the Ivory Coast, the United States stepped in for political reasons with then Vice President Humphrey sent over as personal emissary to show the flag in a former French-dominated area' (Scudder in Eddy and Partridge 1987: 184–210). Such a bold opinion did not appear in a publication in Europe. Where Scudder and Horowitz, the co-founders of the Institute for Development Anthropology, see most impact of their work, Hoben sees USAID's particular failures. Nonetheless, it would be erroneous to use Hoben's conclusion to invalidate Scudder's and Horowitz's claims since the criteria used are not compatible.

40 Sometimes with the public dismissing the other side on normative grounds. For the development anthropologists, see Scudder in Bennett and Bowen (1988: 371) and Horowitz (1994: 3); for anthropologists of development, see Ferguson (1990) and Escobar (1991).

41 In *Encountering Development*, Escobar (1995) addressed development discourse, while his article in *American Ethnologist* (1991) dealt with development anthropology. For an alternative development discourse analysis, see Apthorpe and Gasper (1996), or Achard (1989) using Chomskyan grammar.

42 'To consider culture as texts which say something about something creates the risk of letting them say anything and notably truisms' (Augé 1994: 83, my translation).

43 See Murray in Wulff (1987: 223–240), Escobar (1991, 1995) and Gow (1993). 'The involvement of anthropologists in the AOP has been exceptional: with the possible exception of Vicos, it would be hard to imagine a project in which there has been as active an anthropological presence' (Gow 1993: 384). Another suitable example is the Manantali resettlement scheme in Mali, with the development anthropology exposed by Dolores Koenig (Koenig 1997; Koenig in Chaiken and

Fleuret 1990: 69–83; Scudder in Bennett and Bowen 1988: 379), and anthropology of development explored by Adrian Adams and Jaabe So (1996). Adams and Jaabe So demonstrate that the political exclusion of peasant farmer organizations was reinforced by USAID and that these organizations effectively used NGO funding. They observe that development anthropologists have reintroduced artificial flooding from the Manantali dam in the debate among planners (Adams and Jaabe So 1996: 270), thereby pursuing peasant farmers' interests in the governmental domain.

44 Klitgaard used the AOP as a key success story of USAID to dismiss the insight of anthropology (Klitgaard, July 1991: 95). A shorter version was published later in Serageldin and Taboroff (1994: 75–120).

45 For example, for the prominent French ethnologist Michel Leiris, ethnology was the natural ally of the colonized. The latent processes defined with our case studies suggest that this is still valid.

46 Development anthropologists cannot afford to be simplistic: 'And social scientists can help USAID make more progress in Africa as it works with Africans to reverse the dismal trends and take advantage of the opportunities for better lives' (Hess 1991: 140). After this typical development rhetoric, nothing more than the reminder that anthropology is based on holism shows that little has been learned during 'New Directions'. Providing Social and Institutional Profiles as input into strategic planning (ibid.: 142) can easily be new jargon for what used to be Social Soundness Analysis.

47 Especially in Chad, where the relations between international donors and the government are core political factors, feasible research questions would have been an obstacle. It is important to see the project actors react as freely as possible to the researcher's presence.

3   Constructing the intelligibility of the events based on participant observation

1 'Protagonist' is a label used in French contemporary anthropology to describe an actor who is pursuing an agenda coherent in his/her own life-world. Writing about a group of actors implies that they 'act' on a 'stage', bringing them together. Several protagonists interact equally on a stage, but their view on what constitutes that stage can be quite different. Their interaction can be determined also by the differences in their interpretation of the nature of their relations. In that sense, the individuals we describe are 'protagonists' of the social process that they saw reflected in the projects.

2 With the exception of most Chadians, all individuals mentioned in this book have read this manuscript before publication. Their responses are integrated throughout this analysis.

3 M. Abélès, G. Althabe and M. Augé started their scientific work in Africa in the late 1950s, and then applied their fieldwork skills in France. Today, they work in partnership at the Centre d'Anthropologie des Mondes Contemporains of the Ecole des Hautes Etudes en Sciences Sociales (EHESS), where I wrote the first version of this study. Their recent publications, as well as those translated into English, are included in the bibliography. The Centre's other researchers are J. Bazin, A. Bensa, M. de la Pradelle, J. Friedman, J. Jamin and E. Terray.

4 It is up for discussion whether this approach can provide the core of the interface research pursued by Long. Because of the solid anthropological roots of Long's methodology, this seems to be the case. But for the same reason, the institutional reaction to the results that were obtained is even more difficult to interpret, as

suggested when we referred to Marc Poncelet (p. 21). We return to this in section 3.3.

5 Cogeneration, or Combined Heat and Power (CHP), consists of electricity generation, for example with gas turbines or large piston engines, and waste heat utilization in industrial processes or for residential use. Particularly in the USA, this technology expanded after the second oil shock (when there was a steep rise in the price of oil) because it allowed the overall thermal efficiency to be raised from around 40 per cent up to 80 per cent in the best cases.

6 We will discuss the Mexican myth surrounding Malinche, a symbol of cultural betrayal, below. See note 24, p. 205.

7 The foreigners in Appui Technique, who accepted the economic constraints that led the buyers to insist on the lowest possible prices, were paradoxically liberated by the absence of quality control, which gave them the leverage to redefine their involvement whenever they lost sight of their initial goals: '*c'est quand même pas une raison pour faire n'importe quoi!*'

8 These ideal-types are reinforced by the project design process (the IBRD 'project cycle'). Development agencies produce terms of reference (TOR) that leave the experts a margin of definition of their work, which they use, knowing the agency's political constraints, to justify technologically the disparities between industrialized nations and countries said to be developing. Experts excel in allowing leeway in this definition in order to accommodate the local conditions in the target society that give substance to a project. The developers are then expected to take account of the particularities of the target country in their practice. What the design process excludes subsequently traps the developers during the implementation. '... personal experience, political commitment and technical training ... coalesce to form a specific development discourse in which individuals think ...' (Kaufmann 1997: 129).

9 GRET, the NGO implementing Appui Technique, does little work in France, and so its experts gain their expertise on their missions. This was not the case for Hagler, Bailly, Inc., the consulting company implementing Autogeneración. In the field, the foreigners were able to distance themselves from the institutional discourse. In their home countries, knowledge of Third World reality was an asset, an advantage. In the field, it became a hardship.

10 John had an M.Sc. in mechanical engineering from UCLA (the University of California at Los Angeles), was 38 years old and single. He spent 3 years with an energy consulting company in Boston, then took an assignment for 3 years in El Salvador; he joined Hagler, Bailly, Inc. 9 years before Autogeneración. In the beginning, he was meant to be a technical resource person. After the first few months, the director of the Mexican agency expressed his dissatisfaction with the progress made and John became head of the team of US experts. Until the end of Autogeneración, he remained the dominating foreigner. Today, he manages much larger international energy efficiency programmes, altogether a rather successful engineering career. He has married and has one daughter.

11 Martin was a certified technician and started work in the public sector in France. He remembered having seen high chairs for the administrators during the last colonial days. Most of his career was dedicated to appropriate technology development, first in Zaïre, then for 6 years in Burkina Faso (for the International Labour Organization; ILO) and later in many other West African countries. He supervised and designed projects for two well-known French development NGOs (full of 'soixantehuitards', he said) before joining GRET, where he had designed Appui Technique. Acting as a co-ordinator at GRET's central offices, he spent about 6 months in Chad, assisting the project director, Jacques. His wife and their two daughters had accompanied him for many years, but he went alone to N'Djaména.

12 If technical assistance is a self-sufficient exercise, it is here that the meaninglessness of its practice is being reproduced.

13 Although his family came from a well-known, Northern, nomadic ethnic group, Mohammad grew up in Sarh in Southern Chad. His father, a merchant, was killed at the beginning of the civil war and Mohammad fled with his mother to N'Djaména. At the age of 12 and without formal education, he started to work in a garage. By the time he was 20, he moved to a machine shop and achieved recognition for his craftsmanship after 4 years. At the beginning of Appui Technique, he had been an independent artisan ('un patron') for 2 years. This microenterprise employed six workers and had allowed him to build a large house and bring his extended family back together. He had two wives and no children.

Altogether, forty artisans participated in Appui Technique at various times. The skill acquisition over a long period and finally the big investment of buying one's own machinery was a core thread of most individual life histories of these artisans. Another common element was a period of 1 or 2 years as a refugee in Nigeria.

14 Osama had moved to N'Djaména at the age of 14 from Abéché, in Eastern Chad, looking for professional work. His schooling was better than most other artisans and he could write Arabic and French well. He had also travelled throughout Europe for 6 months. He had decided to stay in Germany, but his money ran out before he could learn German. With his close friend Rahman, he had been running a small workshop for 8 years. Because of their technical competence, they made more profit than most others. He was married and had four children, and his house had the appearance of a modern Islamic middle-class home.

15 He noted with satisfaction that his 5-month-old daughter, who had fallen asleep on a cushion next to me, was not afraid of me even though I was the first White man she had seen. He mimicked my ethnological method and recorded our conversation on his tape recorder with a tape that I had given him. This interview is used again to identify management goals appropriate to the implementation of Appui Technique (section 6.2).

16 Ngerbo, another artisan, said of Osama *'that one is the complicated one'*. For 3 weeks, Ngerbo helped Osama to work on a prototype of a grain mill, without ever reacting to Osama's grumbling.

17 Miguel had an M.Sc. in electrical engineering from a Mexican polytechnic. He had spent 16 years in a utility company (Comisión Federal de Electricidad; CFE) and 4 years with ABB Mexico building a local research unit before joining Autogeneración as the Mexican project leader. He was 45 years old, married and had three children.

18 Ramón had an M.Sc. in electrical engineering from the Universidad Nacional Autónomo de Mexico (UNAM). After graduating, he joined a small consulting firm, run by the son of a former CFE president. He was 27 and single.

19 While Mexicans and foreigners had equal status and nobody could give orders, John, Jim and David could, to a limited extent, use their superior engineering knowledge to force the others to take certain decisions.

20 Miguel had been in the process of writing a fax when I arrived. He held a ruler over the paper to write in straight lines. He had learned this trick from a German engineer, he said. It was one of the few aspects of foreign knowledge that he was able to appropriate. Upon leaving, he invited me to visit his home in such a way that I would not be able to do so. He saw my research efforts as being similar to his efforts to achieve professional independence: *'me gusta mucho como tu tratas de superarte en tu trabajo'*.

21 José was a mechanical engineer and had graduated from the polytechnic. He had worked for 20 years in a large steel mill and had gained a good reputation for his design of plant water treatment and steam systems, which allowed him to work

as an independent consultant. He was 46 years old, married and had three children.

22 This will be explored in the section on communication structures (section 5.4).

23 Having participated in the exchanges between foreign and local actors, I had to ask only simple questions at the beginning of an interview for the interviewee to return to these exchanges and pursue what he or she had previously tried to explain.

24 According to popular Mexican myth, Malinche was so devoted to Hernan Cortès that she betrayed her people. The alliance (a *'realpolitik'*) of several indigenous peoples with the conquistadors against the Aztecs seems to be another matter for contemporary Mexicans. The historical Malinche herself remains a fairly obscure figure, but the mythology of her betrayal has passed into common language as an insult, 'Malinchismo', levelled at someone who has taken the part of the foreigners against the people of Mexico (Paz 1961: 86; Todorov 1984; Cypess 1991; Nunez Beccera 1996). Malinche has also been studied as an historical mother and the source of ethnic identity healing of oppressed people. But since the 1940s, the term has been predominantly pejorative: 'malinchistas – los verdaderos hijos de la Malinche, que es la Chingada en personna', 'the strange permanence of Cortés and La Malinche in the Mexican's imagination and sensibilities reveals that they are something more than historical figures: they are symbols of a secret conflict that we have still not resolved' (Paz 1961: 87). See note 23, p. 218.

25 To avoid changing the project dynamics, I conducted all interviews during the last 10 days that I was present. This precaution was an overreaction caused by my fieldwork stress.

26 'Nasarra' is the local label for White. It is used in many countries in Western Africa and the term comes from the Koran, where it denotes Christians. Nasarra also refers to the White domination during the colonial period. Labelling an act as 'nasarra' (like an adjective) denotes it as being culturally superior and potentially dangerous. Decolonization as a cultural process is reflected in the semantic evolution of the term nasarra.

27 Institut de Recherche pour le Développement (formerly ORSTOM), the French national research institute for developing countries. Working predominantly in former French colonies, it consists of more than 800 researchers.

28 Interestingly, a few days later, Osama refused to tell me the name of the street where he lived. This demonstrates the extent to which the encounter with me remained cultural and was not in any way indicative of real personal interest in me.

29 In Chad, personal contact rendered the lived experience more violent, and the symbolic work (which, thus, became more intensive) more strongly expressed. In Mexico, visits to plants (always by one foreign and one Mexican expert) were rare events. A visit produced enough data for weeks of work for them.

30 To the extent that an encounter between foreign and local actors is an ethnological experiment, the observer responds to an unspoken need and fills an unnamed, but inevitable, position. In this sense, the observer has immediate access to the exchanges and participates wholly in them.

31 Further examples are analysed in section 5.4.

32 Three years before Autogeneración, she had finished her M.Sc. in chemical engineering from UNAM. After 1 frustrating year in a nationalized chemical plant, she had joined a dynamic Mexican consulting company. After Autogeneración, she moved on to another company, where she works today on electric systems in mining. She is still single and wants to get a PhD.

33 I visited her 1 year after the end of Autogeneración in her office in Mexico City. Her comments on the details of the results showed that she had read them carefully. For example, she had found all citations where foreigners talked about

Mexican women. I did not tape this conversation to mark the end of the research. I encouraged her to explain the results to Ramón, but he subsequently avoided meeting me, making one appointment after another. His behaviour could reflect that, during his interviews (where he had explained that he wanted to make money from the gringos), he had misinterpreted his part in Autogeneración. Meeting me again would have meant admitting that his professional rationale had been wrong.

34 Interestingly, Aníbal thought so too, according to the interview we had. I left my interview with him until last, after Autogeneración had ended (see p. 71).

35 Jack grew up in France and was a mechanical engineer. At the time of writing, Jack still worked for the French consulting company who had subcontracted him to Hagler, Bailly, Inc. to work in Mexico. He spends much of his time in former Soviet Union countries and still talks of working part-time in order to stop his constant travelling. He is 39 years old and single.

36 This tolerance was in fact a necessity. Some years later, for example, the IBRD had to decide on whether to exploit the Chadian oil reserves in the absence of a competent local administration. All kinds of development agencies were present and had provided a substantial (and confidential) part of the government's budget. In addition, since the end of the civil war in the mid-1980s, there had been a French fighter plane squadron and 2,000 soldiers stationed permanently in N'Djaména (the capital city). The foreign presence was a precondition for security, social services and a democratic political process; see Azevedo and Nnadozie (1998), Tubiana (1994) or Buijtenhuijs (1993).

37 It should be noted that we cannot deduce from the evidence provided here that a different presentation of my results would not have rendered them more useful to the actors.

38 Agencies have no way of tracking the implementation of projects. We have suggested that agencies function as blind carriers of the sociopolitical agendas encoded into the technical knowledge. It is possible to see the repercussions of this political baggage on the local economy by identifying the motives behind the experts' actions. More importantly, it is possible that local experts act as invisible mediators for an agency, unknowingly determining with whom the agency unwittingly negotiates and therefore the success of projects. The efforts of agencies to understand what happens in technical assistance are labelled as evaluations. These will be scrutinized in detail after the completion of the analysis (section 5.5).

39 The head of one of GRET's offices later conducted an evaluation in N'Djaména that used methodologically similar tools, such as unstructured interviews.

40 This journal is widely distributed in West Africa. The artisans that participated in Appui Technique were doubtless in complete agreement with the last line of the article: 'It is a matter first and foremost of being more like partners than instructors'. They would use this institutionally biased material to confront the foreigners with the contradiction between their behaviour and the profile of the developer explained within. Martin did not consider this when he sent me the journal with a letter of thanks. The strongest misrepresentation in the article is taking the significance of the technical knowledge (oxcarts and grain mills) at face value. This is another fundamental principle of the technical assistance trade.

41 The kind of 'othering', production of cultural distance, useful in the French context.

42 GTZ is a German governmental development agency. Since Moser-Schmitt's study, German anthropologists have tried to influence the policies of the GTZ in much the same way as US anthropologists have on occasion influenced the IBRD (see Bliss and Schönhuth 1990; Schönhuth and Kievelitz 1991; Bliss and Neumann 1996; Schönhuth 1998). They succeeded in providing analyses of the target

population in those areas which were particularly problematic, namely in rural development and delocalization owing to public works (dams, highways, etc.). Because of these fields, it was difficult for them to move beyond the developmental logic of these projects. When Moser-Schmitt returned years later with her graduate students on a field trip, she found that her results were only used in the promotional material produced for the project.

43 Being an engineer and an ethnologist could have avoided some turf battles between engineers and social scientists during my participation in the two projects.

44 Technocratic development agencies declare the need to ascribe greater importance to technology (developmental objects), thus reducing the social dimensions of technical assistance in their discourse. Less technocratic agencies permit a fuller interrogation of the status of technology.

45 Despite their exploitative use of the results, their gesture introduced a practical perspective on technical assistance into the institutional discourse. The article added to GRET's knowledge of Appui Technique, notably regarding the pedagogy of vocational training. Finally, these conclusions confirmed that being paid by the US consulting company did not affect the response to the results (although I received no compensation from the French NGO running Appui Technique).

46 Given the limited space, we cannot return to the differences between development agencies and scientific practices in different countries.

47 This is the conclusion that a famous philosopher drew from his theorizing on the role of industrial technology. Modern society would be so dependent on technology that only a new spiritual force could compensate for this dependence. Constructivist approaches to technology show, on the contrary, that science and technology are only one among many arenas where social and historical processes are at work. Key references for constructivist approaches to technology are Pinch *et al.* (1987) and Latour (1987). For a new initiative on combining technological change and international relations, see Talalay *et al.* (1997).

48 Through their financial assistance, the IBRD and the International Monetary Fund (IMF) impose policy conditions on developing countries. But technical assistance (TA) is not strongly linked to financial assistance. Therefore, an analysis of TA as a viable objective does not imply a normative position to judge power in international relations.

4 Interpretation of the events

1 Participant observation is the methodological basis.

2 The only aspect common to the projects was that the economic feasibilities of the technologies involved were to be assessed and documented.

3 The difference between Hagler, Bailly, Inc. in Washington and GRET in Paris is as great as that between the clients of Autogeneración and Appui Technique. The organizational conditions of a US consulting company are distinct and fundamentally different from a French NGO. In France, careers are relative to a French scale of professional qualification with little direct feedback from day-to-day performance, whereas in the USA a consulting company is based on the principle that anything goes as long as the client is satisfied. These differences cannot be related to the differences in behaviour during project implementation.

4 The word Other (upper case 'O') refers to the constructed identity that each side projected onto its colleagues across the interface. The other (lower case 'o') refers simply to the other group itself – to the opposite side of the interface. Other is the corresponding opposite of identity, separated by alterity (otherness is another term for alterity). Anthropology has produced much evidence for a social Other, of a different nature from an *alter ego* in psychology. I use the term

Other only in the sense of a social Other. Octavio Paz uses the following as a foreword for his poetic masterpiece:

'The *other* does not exist: this is rational faith, the incurable belief of human reason. Identity = reality, as if, in the end, everything must necessarily and absolutely be one and the same. But the *other* refuses to disappear; it subsists, it persists; it is the hard bone on which reason breaks its teeth. Abel Martín, with a poetic faith as human as rational faith, believed in the *other*, in "the essential Heterogeneity of being", in what might be called the incurable otherness from which oneness must always suffer' (Paz 1961). For many anthropologists, alterity is the original object of their discipline.

5   Since my aim here is to reconstruct the day-to-day experience of the actors, I too must leave the term 'project' undefined until the end of the study, when its meaning for each group has been constituted through examples.

6   Domination refers here to a process of normalization rather than oppression, one that is maintained by the regulation of knowledge rather than its suppression.

7   There were approximately 400 metal workshops spread out over N'Djaména. Altogether around eighty have, at least, sent someone to have a look at what was happening in Appui Technique. For several weeks, Jacques and Tahem had been distributing invitations with a sketch of the location to announce their activities.

8   Notably those who employed more than ten workers. The most successful and dynamic artisan could participate only once a week, but he lent tools and machines to the experts and other artisans, showing his commitment to their collaboration.

9   Reasons for this limitation included the fact that the Chadian experts were less aware of the constraints on the normal working conditions than the artisans. University educated, the local experts were also predisposed to sympathizing with the foreigners. However, these sources of discomfort could have been overcome if there had not been a serious lack of understanding on both sides of the exchanges.

10 Although the French experts were well aware of the difficulties the artisans faced in coming to work on the prototypes (they were forced to shut down their own workshops during the hours that they spent at Appui Technique), this acknowledgement did not enter into their exchanges.

11 They did not go so far as to believe that they needed to explain the local conditions to the West Africans.

12 Most of these reproductions did not find a buyer. It is possible that offering these machines to one of the importers of that type of machine would have succeeded. While the artisans were good craftsmen, their marketing was rudimentary. The final evaluation by GRET that they had only been able to create additional sales of $US140,000 remains correct; compare this with the total cost of Appui Technique of $US1,000,000.

13 In fact, this was not the case, but these experts were not the only ones who reached this erroneous conclusion.

14 French Ministry of Foreign Affairs

15 No individual was charged with determining the end of the work day. It was spontaneously decided by the group each day.

16 In section 5.1, we show that experts and artisans shared the same criteria for quality, the disagreement only concerned the ways to guarantee quality. Many artisans worried about this also because they saw that other artisans did not have the same craftsmanship as their own.

17 Describing these other aspects, we follow the actors' efforts of interpretation, which advanced by completing the puzzle of their encounter.

18 Mohammad played a similar role among the artisans.

19 Furthermore, his perspective on the Chadians was surprisingly more

developmentally conscious than that of Martin or Jacques, who both attempted in their behaviour to distinguish themselves from traditional developers.

20 We will return to the importance of this attitude below.

21 Martin proposed this approach verbally in meetings between the experts, but would never discuss it in front of the artisans.

22 A marvellous act of portraying himself as exotic, anticipating an adequate level of misunderstanding between himself and me.

23 I have already introduced María, José, Miguel, Ramón, John and Jack. Ben, Jim, Joe, Bill and Eva appear later. Tom is the author. The remaining Mexican actors are not introduced individually. The foreigners were of Colombian, Argentinian, Cuban, Peruvian, US American, French and German origin.

24 These plants belonged to companies whose annual turnover was larger than the GDP of Chad. Indeed, the industrial context in Mexico represented the opposite end of industrialization.

25 Hagler, Bailly, Inc. refused to participate when Autogeneración was to be repeated. The Mexican Agency currently uses their consulting services for other activities in this field.

26 John claimed that better Mexican engineers were available, but that they had more important things to do. Jim felt that the Mexican Energy Agency had acted in bad faith.

27 Ben was the oldest foreigner at 49. He received his M.Sc. in Havana, Cuba, from where his family emigrated because his father, a Protestant minister, could not continue to work. He had spent 15 years in a US utility company. A specialist in boilers in sugar mills, he was also a consultant to different Latin American sugar companies.

28 Jim first studied in his native Argentina before moving to the USA to complete an M.Sc. in mechanical engineering. He gained a reputation as a project manager in hydropower studies with a Canadian firm. Montreal was 'the place to be' for him, but he had to earn the college fees for his three children and he earned a higher salary in the USA. After Autogeneración, he directed energy studies in Eastern Europe before returning to a quieter life in Canada.

29 Joe studied in Lima at the military academy and had been working as an energy consultant for 3 years before moving to Oklahoma City. He had directed research in industry in the USA and was an assistant professor at a university. He was married and had five children. Hagler, Bailly, Inc. staff called him 'the teacher'.

30 For example, by promising salary benefits that he knew were not available in Autogeneración's budget.

31 Finishing his engineering degrees, Bill joined the US Navy and worked on board ships. Settling down to raise a family, he worked for 10 years for a power plant construction company, where he had been a costing engineer (calculating investment costs) for renewable energy plants.

32 When I entered the office after 2 months of absence, he greeted me with almost hysteric laughter: '*Hey Tom, I'm so glad that you're back, nothing has changed, they have no confidence in us, it's all up in the air!* '

33 For example, the firm conviction that he converses with God and the aggressive appreciation of feminine elegance symbolically allow the Mexican to distance himself from the image of the poor passive native.

34 The Mexican engineering companies that the Mexican participants worked for were known to be in competition for business. To their credit, the Mexican experts acted in collusion in the presence of foreigners, avoiding disagreement. This was not difficult for them as their contracts with their various employers were actually drawn up at a time when they all had one and the same client, the national electricity company. Their common background was an evident source of solidarity for the Mexican experts. For the latter, this kind of competition signalled,

alternatively, a lack of professionalism (María) or the desire to keep some aspect of their world hidden from the foreigners' view (Miguel). The foreigners, for their part, did not try to exploit the competition. They encouraged their Mexican colleagues in their work regardless of their affiliations. Similarly, the Mexicans never discussed their employers with the foreign engineers. The circumstances of the encounter with the foreigners prevailed over the influence of their professional past.

35 When the foreigners' superior came to Mexico to visit the Agency, he would not do so in my company. John considered this with irony, saying, *'He simply doesn't want you to hear him bullshitting around!'* This ostensible critique of his boss was in fact John's way of suggesting that 'bullshit' was the discourse of choice for US experts in their dealings with Mexicans. In this sense, then, the behaviour of his boss confirmed and justified the foreigners' prejudices.

36 The reader will recall also that the presence of a third party, the local experts, at Appui Technique helped to create an interface with multiple discreet levels of exchange.

37 Described on p. 96.

5 Latent processes in technical assistance

1 Both the unperceived disagreement on quality control and the perceived difference in defining technical acumen only represent the risks and benefits of the encounter of these individuals, which could easily have also produced quite different results.

2 The obstacles to organizational learning can be defined by confronting the implementation with the discourse in development agencies. This would be specific to the involved institutions, i.e. the IBRD and the local governments.

3 Which does not imply that a sociology of knowledge would not be able to identify differences between local and foreign knowledge. My point is that these differences cannot explain the encounter between these actors. Differences between local and foreign knowledge are probably less influential for developer–developee encounters oriented towards know-how, i.e. the transfer of know-how between local and foreign experts through demonstration applications. In such cases, the implicit application-related content, the know-how of the expert which cannot be explicitly isolated or taught, is anticipated by the actors.

4 Chad exports cotton and imports the oxcarts necessary to produce the cotton (about 5,000 annually). Cotton being the basis of the local economy, a public company distributes oxcarts to farmers in Chad's cotton belt, often in exchange for part of the crop. Farmers sometimes come to N'Djaména looking for an artisan they heard about in their village to get their oxcarts repaired. Chadian cotton is of the highest quality, but low-cost and poor-quality producers in the USA imposed low and rigid quotas for Chadian exports through the famous Multi-fibre Agreement, a part of the GATT (General Agreements on Tariffs and Trade) rounds.

5 The tools and the raw material for the prototypes were paid for from the Appui Technique budget. All of the actors at Appui Technique could solder, drill and bolt together pieces of iron, although their individual dexterity and skill varied as well. The reputation of an artisan lay in the products of his labour, e.g. the successful transformation of wheat to flour in a hammer mill.

6 This hypothesis was, in fact, confirmed by both the foreigners and the locals of Autogeneración.

7 Knowledge referring to knowing how to swim, for example, compared with know-how referring to the capacity to actually swim.

8 These are famous case studies by Wiebe Bijker and Donald MacKenzie. 'Seamless' is the jargon, implying that one can study the structure of the fabric. However,

wanting to separate technology and culture actually only renders the identification of the fabric's structure more difficult.

9 Import substitution as a stage of industrial policy for late industrializing economies.

10 He was the only artisan who refused to tape an interview: *'you ought not steal my words'*.

11 See, for example, McNaughton (1988).

12 He asked me to procure for him sorcery perfumes after I had spent 2 weeks in his workshop welding tables and chairs. After 15 years of civil war, it is no wonder that there are many practices of sorcery producing magic shields against bullets. Welders have special status when they have worked on battle tanks. Mohammad told me about some who tried to manufacture a tank, where the prototype's engine burnt out. I hope he did not do that because he knew of my German nationality.

13 The *One-dimensional Man* is the title of one of Marcuse's influential *chefs-d'oeuvres* of the 1960s. To my knowledge, technical assistance practice has never been dealt with using the results of theoretical work on technology. Marcuse seems to be right on target, although his conclusions have been questioned. Regarding developing countries, this has been exposed in a volume produced by Jürgen Habermas's team, *Antworten auf Herbert Marcuse* (Habermas 1968), especially by Heide Berndt. At that time, Habermas elaborated on his *Science and Technology as Ideology*, published in honour of Marcuse's seventieth birthday (Habermas 1970).

14 See the debate on development anthropology and an anthropology of development in section 2.4.

15 This theorization of global cultural and social processes is best presented in Ekholm and Friedman (1995) and Friedman (1994). They use Bourdieu's concept of *habitus* to describe these practices of building social identity. A habitus is a whole set of practices that an individual uses to construct meaning in his/her life. A habitus consists of habits of working, loving, eating, etc.; one habitus contains a script for all facets of human existence.

16 Ekholm and Friedman (1995: 162). Atmanspacher and Dalenoort (1994) provide an overview of exo- versus endo-concepts in natural and social sciences. They show that exo-concepts follow from a theoretical position and do not exclude endo-concepts. In fact, speaking of an exo-social process implies that an analysis with outside (exo-) concepts is more pertinent than an analysis with inside (endo-) concepts. Both are always possible. An exo-social process is intelligible from outside, reducing the significance of inside conditions. Epistemologists, especially in cognitive science, experiment with these concepts to avoid Cartesian dualisms. For an exploration of a sociality among knowledge experts, see Knorr Cetina (1997).

17 This diminishing and even destruction of meaning was literally enacted by Osama breaking numerous drills. He also used the labels *'comme en France'* and *'comme au Chad'* more than anybody else. Other expressions of the destruction of meaning were the many instances of unnecessary effort, e.g. sawing steel rods (quite tiresome by hand) which everybody knew were not to be used.

18 This conceptualization is not functionalist, however, because it is not concerned with the different outcomes between endo- and exo-social processes.

19 The experience in Mexico was of less value to the foreigner's career than a project elsewhere in the world. For the Mexicans, their presence was an indictment of their nation's technological achievements and capacities.

20 This term is used in the development literature to describe examples of misguided 'improvement', such as the installation of cardiac units in areas that only practice simple medicine or of sophisticated computer systems in universities whose libraries are sorely lacking in printed material.

21 The notable exception to this tendency to reinforce cultural distance was Mohammad.

22 We have noted that the experts carried with them the education and the scars of their past work in technical assistance. In both of these projects, the most pragmatic actors, those who carried the fewest scars into battle, were the most likely to function effectively. Unfortunately, their efforts were rarely visible.

23 One case constitutes an interaction of two exo-processes, Chadians and French maintaining their perspective on the other; the other case constitutes two endo-processes, the US Americans and the Mexicans struggling with their perspectives on themselves. It is also possible to deduce interface characteristics from this suggestion.

24 Ekholm and Friedman (1995: 164) assume that the transition from exo-sociality to endo-sociality occurred in Hawaii at the end of the nineteenth century.

25 Stocking (1991), Breckenridge (1993) and Mbembe (forthcoming).

26 Although Mexico is a member of the OECD today, and the stocks of its biggest companies are listed on the New York Stock Exchange (Telefonos can be the most traded stock), it is less affected by world politics than Chad, a country that knows no foreign investment. Chad is much more often than Mexico the site of global confrontation. In Chad, Somalia or Sudan, the relationships between the USA and the former Soviet Union and between France and the Arab nations have profoundly influenced the political climate. Mexico is better insulated from exterior influences. Furthermore, Mexico's independence is as old as that of the USA, while Chad is still dependent on France.

27 The scientific context is in section 2.3.

28 Marc Poncelet (1994: 147) suggests that this way of looking at the issue is particular to rural sociology.

29 Unemployment would be another unwanted macrophenomenon.

30 'Interface encounters do not remain constant across all social contexts for the actor involved. An actor-oriented approach therefore must try to identify the conditions under which particular "definitions to reality" are upheld and to analyse the interplay of cultural and ideological oppositions' (Long, 1989: 239). Long, as well as Friedman, uses the notion of *habitus* in Bourdieu's theory of practice to define the reality one actor pursues in diverse situations.

31 Student, ethnographer, engineer, idealist, atheist, German and so on.

32 This popular saying is often attributed to General Porfirio Díaz, President of Mexico before the Revolution.

33 Private power implies the decentralized generation of electricity in combination with industrial production. Because of this combination, private power is often more efficient than centralized utility power generation. Private power development has also expanded considerably because many developing countries do not have the financial resources to expand electricity production at the necessary pace to support economic growth in general. In some countries with severe energy shortages, private European and US American companies have taken control of this sector in recent years. Private power development also has an important part to play in climate change mitigation, a challenge to which many energy companies do not rise.

34 Eva worked for 1 year as a secretary, typing the feasibility studies.

35 He was one of four experts who never, in the office, alluded to my research as John and Ramón did. Being more aware of the differences in interpreting the collaboration, he probably saw no opportunity to let my participant observation appear to be useful to this collaboration. Carlos was also the only Mexican expert resenting, in particular, Miguel's insistence on his individual contribution.

36 José pointed to one such serious misunderstanding at the very end of his eulogy during the end-of-project dinner, '*we are all proud to have participated, but one thing I want to say to you, John, your vision of Mexicans is erroneous. We are not as weak as we pretend to be*'. John replied with a smile. That moment was the symbolic end of

Autogeneración. The foreigners did not understand this and nothing more was said about the project. José had been the only Mexican to speak.

37 See p. 64.

38 Today, these prototypes are still gathering dust in a shed. After having turned once, nobody had any use for them.

39 He anticipated why the foreigner's strategy to compete with the imported oxcarts would fail. The artisans could offer oxcarts cheaper as they only imported the ball bearings and the wheels. Martin had already calculated that the artisans would conquer 60 per cent of the local market, 3,000 oxcarts annually, and had estimated how many Chadian jobs that implied. Dambai knew already why Martin's quantitative analysis was doomed, but did not explain it to Martin. Possibly because he had no solution.

40 Two days before I was due to leave Chad, he had invited me to dinner at his house and we talked for 3 hours.

41 The other experts did not perceive his subtle efforts. In fact, Atula, the Chadian expert, and Martin, the French expert, both thought that Dambai was afraid of the artisans.

42 A remark during a meeting of the experts. They met daily at 7 a.m. Whenever I did not attend, I asked one of them to start my tape recorder at the beginning of the meeting. The artisans arrived at 8 a.m.

43 Austin proposed the term 'illocutionary act' to analyse the meaning of speech acts and to grasp the nature of the action, or the act of producing an utterance. The term 'illocutionary' is often used to express the actors' mutual agreement to engage in an interpersonal relationship (Habermas 1988: 380). The illocutionary content of an utterance corresponds to the most subtle mechanics of power. 'In illocutionary acts, the symbolic efficacy of an expression is a function of its capacity to valorize its capital of authority and its argumentative capacity to seize the constitutive legitimacy of the field by announcing itself "authorized"' (Habermas 1992: 83). By according an essential place to instituted illocutionary acts in his analysis of performatives, Austin sent a wake-up call to sociologists and linguists alike. The richness of the discussion on the use of the term 'illocutionary' demonstrates the importance of these questions for research.

44 We use the distinction of three types of communicative action but do not follow the theory as far as performing a linguistic analysis. For Habermas, the illocutionary types are developed into types of interaction to permit a fusion of philosophical and sociological theories of social activity. Our reductive adaptation does not correspond to a strict application and is meant only to clarify the communication process at work in technical assistance. Bierschenk (1990: 21) and Grosser (ORSTOM 1986: 313) have also isolated and used discreet Habermasian concepts in their work: the distinction between communication and strategic acts and the relationship between dramaturgy and strategy respectively. Our adaptation is more limited because we do not use the theory to say something about the purposes or the actual outcomes of the exchanges. We qualify only the form of the exchanges.

45 Eva worked for 1 year as a secretary, typing the feasibility studies. John often asked her to work overtime because he felt she was particularly ambitious. She appreciated this, notably because she hoped to demand a higher salary in her next job, having worked for foreign experts. She often reacted emotionally and she was the only Mexican to take pictures in the office, which she later sent to me.

46 Mead taught at the University of Chicago from 1894 to 1931. Habermas's 'theory of communicative action' is built on Mead's oeuvre (next to Max Weber and Talcott Parsons). In the second volume (Habermas 1984), Habermas uses Mead to establish that language is the form and the basis for human socialization. Often, commentators on the theory of communicative action assert that

Habermas corrects Mead with respect to the autonomy of humans in society. Where Mead postulated that humans can detach themselves from society and its values, Habermas incorporated absolutely that the individual remains nonetheless bound by social recognition.

47 The illocutionary force of their acts was greatly reduced.

48 We remarked above that Rahman was chastised for talking and 'opening up' to the nasarra.

49 This kind of confusion occurred on occasion but could also be exploited deliberately for amusement. For example, I asked for information concerning some soldiers who had been arrested some days before. Jacques responded: '*They dressed as customs officials, which they weren't. They were collected and taken to the border in the North so that they could control the importation of camels.*'

Atula and Mondai broke into laughter and Dambai commented: '*Jacques, he has become Chadian, really!*'

50 A term so far from reality that it should be abandoned. Using a term semantically closer to actual practice would by itself be helpful – I suggest 'evidence reports'. For a comprehensive account of the current state-of-the-art evaluation science, see Rebien (1996). He identifies two distinct types, conventional aid evaluation – represented by OECD's Development Assistance Committee (DAC) – and a non-established type – called participatory evaluations.

51 María concluded that John and the other US experts had manipulated the project and dominated the Mexican experts in a manner not consistent with the US experts' interests. Jack, the foreign expert, concluded correspondingly (see p. 45).

52 Of course, the purposes of a technology can be identified and modified, but it would not be meaningful to distinguish between technical and cultural purposes, these remain intertwined.

53 'This is so because people who have a low self-concept and expect failure apparently feel some discomfort when they suddenly perform well, as psychologists have shown. In this manner, social psychology provides a clue to a Latin American phenomenon that has long puzzled me, yet has struck me with such force, that I have invented a name for it – the failure complex or fracasomania' (Hirschman 1987: 187). During his fieldwork in IBRD-funded development projects 20 years earlier, he visited project teams for short periods of time and was paid by the IBRD, two conditions which made him an outsider to the developer–developee encounters he observed. His regional focus as a development expert has also been in Latin America (Hirschman 1965, 1976). Hirschman did his best to challenge the widespread confidence in scientific planning in the 1960s (Hirschman 1995: 127), but suspected that there were deeper structural reasons.

54 'Fourth generation evaluations' or 'participatory evaluations', which appeared over the last 10 years, will not be used to comment on the experiment. The potential of participatory evaluation reflects the strength of the TA practice. Joint evaluations by developers and developees can overcome the limits of earlier evaluations. This study can only provide more evidence as to why participatory evaluation seems to be the *only* feasible approach to improve evaluation. The difficult and slow introduction of participant evaluation indicates that the closed character of TA practice is weakened and that development agencies hesitate to engage in such reality checks. How would John, Jack, María and José go about evaluating Autogeneración? This is a suitable research question to yield methodological tools for participant evaluation. For our purposes, the earlier evaluation mountain serves to verify the thought experiment.

55 The two case studies were designed at the end of the 1980s and Forss's work therefore also reflects TA thinking at that time. No major evaluation effort on TA has been undertaken in the 10 years since Forss's study. This reflects first of all methodological difficulties.

56 '... in northern European countries, the evaluation–function in aid agencies and/ or ministries is better developed than in southern European countries. But also in northern donors' evaluation work some easy improvements could be made' (Hoebink 1995: 9).

57 Counterpart refers to foreign and local experts working together, in order to facilitate a transfer of accumulated know-how. The underlining in this quotation and in those that follow appears in the original text.

58 Eight years later, in 1995, FINNIDA commissioned a synthesis evaluation for the period 1988–95 (Kaponen and Mattila-Wiro 1996). This study still concluded that 'TAP were found to be problematic in many respects', but no suggestions for improvements were made. Regarding evaluations, the study concludes: 'it is important to raise their quality by the following means, among others: more resources, new guidelines, increasing involvement of the partners from developing countries, more social and cultural expertise, and more broader thematic and other such evaluations as well as ex-post evaluations' (ibid.: 7).

59 According to Carlsson *et al.* (1994: 180), this is also a systemic phenomenon in the IBRD: '... the available alternatives are narrowed down not by reference to sophisticated analysis, but through the use of experience and knowledge. It is more a matter of judging the viability of a project on the basis of accumulated experience. Good ideas grow out of personal experience, knowledge and initiative, rather than from a fixed methodology of project selection and appraisal.'

60 Rebien concludes similarly: 'The Grameen Bank follows a participatory strategy not only rhetorically but also in practice, a strategy which is a guiding principle for the entire organization and all its activities. It is unlikely that the participatory loan evaluation system would be working if a participatory strategy were not also behind loan appraisal, group formation, the 16 decisions, the weekly meetings, etc.' (Rebien 1996: 137).

61 This research orientation is still being pursued, see Carlsson *et al.* (1997).

62 On occasion, one can observe that less experienced foreign experts are more successful because they have less accumulated know-how. Approaching the technical content in a less refined manner helps local experts to concentrate on the instrumental core.

63 We started with the assumption that evaluations have become an element of the reproduction of TA failure. Forss would not confirm this because he asserted that evaluations and debriefings have no repercussions; the *ad hoc* solutions to the problems of delivery would not be informed by past experience. This does not contradict the assumption because it invites two comments: first, the *ad hoc* solutions can be reproduced by word of mouth about what is not written in evaluations, and, second, no repercussion is by itself a condition for the continuing of the *ad hoc* solutions.

64 There are studies by academic researchers among returned experts (Elwert on German volunteers, Guth (1982) on French *coopérants*, and some others), but for different reasons these are not comparable.

65 'Technical assistance' in *Lessons & Practices* no. 7 was produced in May 1996. *Lessons & Practices* is produced by the Operations Evaluations Department of the IBRD to help disseminate the IBRD's development experience.

For an account of the Operations Evaluation Department, see Carlsson *et al.* (1994: 147–148). The re-estimation of the rates of return in the OED audits after project completion and the use of a cut-off rate of return of 10 per cent seem to serve as legitimizing a certain volume of lending. Baré also stresses that OED evaluations serve other policy goals, citing the famous Wapenhans Report: '[evaluations] were of little value as "lessons learned" as they are rarely read, except for people wanting to know how to write one. (Baré 1998: 322).

66 Ten years lay between Forss's work and the cited OED reports. This is a rather long period in TA and project designs evolve constantly. Regarding the insight

into TA, however, OED lags behind. Carlsson's and Forss's current work on adapting the evaluation function to a development agency (Carlsson *et al.* 1997; Forss and Carlsson 1997) is well ahead of OED.

67 *OED Précis* no. 71: 'World Bank Relations with Mexico' (OED 1994). A précis is a publicly available excerpt of an internal document, in this case of the 'OED Study of Bank/Mexico', report no. 12923. Three-quarters of 163 loans amounting to $US21.2 billion went to specific projects, almost all of which had a TA component, accounting for more than 7 per cent of that funding (ibid.: 4). Over 45 years, $US1,100 million was thus spent on IBRD TA to Mexico. Assuming that expert salaries account for two-thirds of TA costs and assuming an average daily rate of $US500, this should have bought in the order of 6,000 expert man–years. The IBRD still records an exceptional reliance on technology imports and ceased industrial technology financing (see p. 172).

68 Nowadays, cost–benefit ratios are replaced by the economic and/or financial rate of return. For a critique of this practice, see Carlsson *et al.* (1994: 189–203). Their conclusion was that economic analysis is not able to capture the reality of developing countries, but that it enables individuals and groups within an agency to mediate internal and external pressures and conflicts.

69 Which benefits the company operating the plant as well as the national economy because Mexico would export the corresponding amount of fuel.

70 For example Rebien (1996: 5), Scudder in OED (1995a: 224) and Carlsson *et al.* (1994: 4).

71 In that respect, the two case studies are illustrative, the funding agencies and the implementing agencies seek to evaluate but fail to produce lessons to be learned. Notably, the French NGO invested in an evaluation of Appui Technique, including a workshop in N'Djaména involving the project beneficiaries, but it was disregarded by the IBRD.

72 'Direct' refers to the immediate effect of the experts' activities. The structural and the dynamic impacts are secondary, indirect and longer term. Unfortunately, there is no categorization of TA impact types available. If only the direct impact had motivated the design, one would have hired only individual consultants, one for each industrial sector, instead of forming a large team of generalists (with the exception of Ben and José). By this means, the best possible feasibility studies at the lowest cost would have been produced. Nothing in the project documents suggested that the experts were chosen with a particular dynamic impact in mind.

73 What the OED called new conventional wisdom (OED 1996).

6 Technical assistance event management

1 Management manuals often refer to the technical content or to the general context of TA. But neither the content nor the context can predict the dynamics of implementation. Such manuals are imposed by contractual agreements with the actors. Otherwise, actors would not use them, knowing that these manuals do not account for the dynamics of project implementation.

2 In addition, foreign and local experts reading this book can reconstruct the latent processes. We aimed from the start at the actors' understanding of an event, because attaining their understanding of an event alters that event *at the same time*.

3 For example, see Scott-Stevens (1987), Forss *et al.* (1988) or Porter (1991: 98).

4 See Rosenberg (1982).

5 See Keynes (1936: 162) for entrepreneurs' animal spirits; for the culturization of obstacles to technical assistance, see Rist (1994) and Poncelet (1994) regarding the 'Clercs de l'Humanité'.

6 *Development Projects Observed* (Hirschman 1967).

7 Strikingly, both schools of thought remain marginal but influential.

8 USAID was the major client of Hagler, Bailly, Inc. (under Congress's appropriations pickiness), and thus the company grew by internalizing the technological traits of US governments, whereas GRET's major client was the French *Ministère de la Coopération* and 'rue de Monsieur's' habit of 'mise à disposition du coopérant' is a version of French assimilationism. Their experts adapt their management so that the outcomes fulfil political requirements.

9 They are analytically inseparable from my position as an observer, and their wider purchase depends on the quality of the interpretation. Reducing the complexity of the events through the interpretation provides insights but not necessarily universal concepts. For the quality of the interpretation, the change of the interpretation between Chapters 3 and 4 and Chapter 5 is the most important factor. The interpretations in Chapters 3 and 4 are constructed from the point of view of the fieldwork, juxtaposing the cases, pointing to the coherence of the actors' perspectives and integrating the actor's comments on the analysis. Theory is introduced in Chapter 5 and the comments of the interpretation are no longer sufficient to control the interpretations.

10 In the literature, 'alterity' and 'otherness' refer to cultural distance. 'Alterity' denotes cognitive behaviour. An individual's feeling of cultural difference is subjective and the product of that individual. Until now, we have concentrated on the actors' practices of producing cultural distance. Treating cultural distance as an object of analysis would have reinforced this. To avoid this, we have only reported it.

11 The word 'Other' refers to the constructed identity that each side projected onto its colleagues across the interface. The 'other' (small 'o') refers simply to the other group itself; to the opposite side of the interface. In the case studies, this explanation was useful to describe the events. Other is the corresponding opposite of identity, separated by cultural distance. I use the term Other only in the sense of a social Other. Octavio Paz uses the following as a foreword for his poetic masterpiece: 'The *other* does not exist: this is rational faith, the incurable belief of human reason. Identity = reality, as if, in the end, everything must necessarily and absolutely be one and the same. But the *other* refuses to disappear; it subsists, it persists; it is the hard bone on which reason breaks its teeth. Abel Martín, with a poetic faith as human as rational faith, believed in the *other*, in 'the essential Heterogeneity of being', in what might be called the incurable otherness from which oneness must always suffer' (Paz 1961).

12 It is theoretically possible to maintain a neutral position on either or both axes – to remain indifferent towards the Other and his/her character. This rarely happens, however, given the ideological violence of the development encounter.

13 Todorov's epistemological plane.

14 Economic exploitation by businesses in the oldest hubs of colonial power is a function of capitalism. However, this is much more striking in Chad than in Mexico. The majority of foreign settlements in Chad have always been French, whereas, in Mexico, US investments represent only about 30 per cent of the total foreign investment.

15 The EU later financed the cotton mills.

16 Yves Rabier's (1993) doctorate reconstituted the evolution of French foreign policy and concluded that relations with Chad are a French passion, especially in the Gaullist party. 'Chad was the first French territory to resist the German occupation', while the other colonies complied with the Vichy Regime. Général Ingold (1945) described how the later Général Leclerc fought Italian and German troops in Northern Africa from the capital of Chad, then called Fort Lamy.

17 Interview with Ngerbo.

18 Ngerbo: '*On doit arrêter les conneries ici au Chad!*'

19 Rahman understood the risk of being reproached that he ran in attempting to produce models like the White man because he had been told that he was not up to the task, but he was willing to take the risk. Mohammad invited me to share in the local knowledge. He explained magic to me and requested that I help him to obtain what he needed to perform certain marvellous deeds. He spoke, in particular, of magic armour plating for the Chadian military.

20 Ngerbo appealed to the foreigners' authority to judge better than Osama. At the end of the training programme, he alluded to the reputation of the foreigners in order to convince them to give him more concrete help – a stratagem that Osama was not willing to use. Ngerbo's attitude also led him to favour Pascal's pedagogy over Martin's simpler method. Osama did not agree with these differences, although he recognized them.

21 Ngerbo succeeded best because his was closest to their perspective. He remarked in passing that a foreigner in Chad was called an expatriate, whereas a foreigner in France was an immigrant. His sensitivity to categories during the various discourses made him the only actor able to observe the project's evolution with critical distance.

22 Evidently, the artisans were unable to take Rahman's lead and understand, as Rahman was able to, the foreigners' proposition of aid and assistance, and appreciate their intentions.

23 'From childhood on, Mexicans learn to regard that country as otherness. This otherness is at once inseparable from us and yet radically and essentially foreign. In Northern Mexico, the phrase "the other side" [el otro lado] is used to speak of the USA. The "other side" is a geographical reality: the border; a historical reality: another civilization, another language, and above all, another time (the USA is a modern culture while we are still struggling with our past). It is also a metaphorical otherness, for the USA is the image of everything we are not. It is otherness itself, except that we are doomed to live with that otherness, the other side is the contiguous side. The USA is always present in our midst, even when ignoring or turning its back on us; its shadow falls on the whole continent.' (Paz 1972: 67). There is a considerable and varied literature on the gringo. Anthropologists have studied this figure in many different regions of Latin America. Often, they find that the term includes foreign, Spanish, European, Western and modern, as well as specifically US conditions.

24 'A technically sophisticated managerial bourgeoisie wrested national political power from a Porfirian state which by virtue of shifting international power relations, its dependence upon foreigners, and its inability to stabilize domestic social relations, had become increasingly incapable of projecting itself as the seat of national hegemony. A defiant nationalism and strong anti-imperialist bourgeois state became the only alternative to political chaos and anticapitalist revolution' (Hagnes 1991: 250).

25 This assistance constituted only 0.1 per cent of Mexico's GDP (1987 statistics).

26 A classic error of culture-blind management literature in general.

27 Jacques during his interview

28 Their mission was really a kind of prolongation of France's military presence.

29 Hence Rahman's response – to want to '*stay straight with you [tenir droit]* '.

30 These reminders were always received with disgust by the foreigners, except Pascal, for whom the domination of the White man was read in a Christian context.

31 Interview.

32 We have already touched on the point that the actors reacted directly to the rhetoric of the other, a rhetoric upon which the figure of the Other was only one influence.

33 The limited usefulness of the three axes to make sense of the foreigners' attitudes

can be explained in part by the various origins of the foreigners, who came from different parts of the world. They met in the context of the project and shared few common references.

34 By virtue of technology's ideological dimension and of the professional socialization intrinsic in the accumulation of context-specific know-how (expertise).

35 Mexico as well as the USA enjoyed the latest technology, after industrialization, and shared a history of social change.

36 France was post-industrial and Chad was agricultural, and Chad suffered from a precarious politico-military situation at the hands of the French.

37 These two cases were certainly opposites regarding the role of the IBRD in their design.

38 As demonstrated for the political imaginary of modern societies by Castoriadis (1987: 175).

39 The studies carried out by teachers who were asked to construct ethnographies of their multicultural classes in the USA yielded significant results. Many development agencies require training for their consultants that includes a workshop on intercultural communication and sensitivity to local custom (forms of politeness in the language, etc.), but their effectiveness is often debated.

40 The IBRD recently distributed small sums ($US60) to thousands of informal microenterprises across Africa because they could find no other effective means of offering support.

41 Martin could not understand how Osama could manage to break a 13-mm drill bit, and tended to rationalize the problem by attributing it to Osama's low potential as an apprentice, which was in total contradiction to everything else Osama did.

42 He frequently used his own tape recorder to record himself.

43 The interviews of Mohammad and Rahman did not inspire them to reconsider the project or the experts. They shared their professional history with me, as detailed descriptions of their apprenticeships from both an economic and a technical perspective. These disclosures were simply an extension of the discussions they were able to have with the other experts. They established communication with the experts of Appui Technique independent of the distance between the nasarra and the artisans.

44 According to Osama, the United Nations' projects were the most abusive and the projects of the USAID were the least. The practice of aid and assistance by France was marked by the excessive number of developers in the field who hinder development.

45 Osama never mentioned the behaviour of the experts to me.

46 The development dialectic is characterized by the impossibility on the part of the local actor to become like the foreign expert, who civilized himself, so to speak, and by the absence of other models of civilization. The local actor lacks a subject on which he/she could exert civilizing control once his/her promotion to the position of expert has taken place. We should not underestimate the power of the structural dissymmetry between the civilizer, or modernizer, and the native. The modern's position and lack of a third party ultimately prevents the native from becoming modern. This contradiction best illustrates the lived experience of Ramón in Autogeneración and Osama in Appui Technique.

47 The dangerous effect of dramaturgical activity and the evidence of strategic activity analysed in Appui Technique corroborate this theory. However, I do not believe that the analysis of these other communication structures necessarily clarifies the examination in this chapter of the operationalization of symbolic acts.

48 However, it is already clear that if the interest of the foreigners includes a

confirmation of their cultural dominance, or if the local actors submit themselves to what they perceive as omnipotence, it cannot occur.

49 One was an ambitious achiever and the other was a modest intellectual, whom even José, the modest Mexican engineer, called '*el francesito* [the little Frenchman]'.

50 In Chad, Martin identified the microenterprises as a reservoir of economic activity without being able to explain how this reservoir came to be and how the various elements that constituted it worked together. The artisans were simply unable to reconstruct this process. Hagler, Bailly, Inc. had proposed the correction of a structural deficiency in the energy sector in Mexico without establishing what institutional factors led to the inefficiency of the system. The Mexican experts could not fill this void of information.

51 Analysing his personal history – his childhood in Abéché, the highschool in N'Djaména living with his uncle, his visit to Europe and so on – would allow him to reconstitute his understanding of the foreigners. But these personal factors were unique for each artisan, and their lives had been quite different. To use a health analogy, we are pursuing an epidemiology not medicine for individual cases.

52 This interest was best expressed by Jack, the French expert in Autogeneración. He was the only foreigner who read this text attentively in the hope of finding clues to the collaborative exchange that would help him to improve his rapport with Mexican colleagues during future projects. We could also have derived management goals by looking at the foreigners, but because I was a foreigner my exchanges with local experts were more salient.

53 The post-colonial power of the developers was their superior technological competence. The case studies reveal that this power position was not diminished and that the reproduction of these encounters was certain, and with it the developmental failure.

54 This is a non-controversial judgement if one does not distinguish between problems of TA objectives and problems of TA delivery. I believe such a distinction is not pertinent. The predominant position is that this distinction is pertinent in some regions but not in others: 'Nowhere is this more evident than when reviewing sequential World Bank projects carried out within the same sector over a period of years. In these cases, one can find management studies and appraisal reports dating back to the 1960's which identify the management and capacity weaknesses constraining institutional development, and which proposed the same TA solutions that are being proposed today. Thus, after decades of assistance including thousands of man–years of inputs, African institutions are still considered by the donor community to require as much and sometimes more, expert advice and services as was needed just after independence' (IBRD 1994: 7).

55 Today's white elephants are a new species that we are learning to recognize. In fact, we have to look for many such species.

56 Argyris (1978), Schein (1993) and Senge (1990).

57 Rondinelli's conclusion from decades of research of development administration and management that management is 'neither a science nor an art, but a craft' reflects the difficulty of understanding implementation. Development management can be described explicitly, there is no necessity to assume that there are craft capabilities. Management in intercultural encounters requires having the right analytical concepts.

58 We have already cited evaluations from IBRD, USAID and of Scandinavian donors; to complete this, we now use the Inter-American Development Bank.

59 Filling gaps in local institutions is typically done with long-term foreign experts.

60 'This module proposes several solutions to the problem, including (a) involving institutions rather than individuals as counterparts, (b) prudently estimating the availability of local counterparts, (c) encouraging local staff to volunteer for counterpart assignments, (d) clearly defining counterpart qualifications, and (e)

involving counterparts in the coordinating process between the Bank and project management' (IBRD 1993: 103).

61 'It is understood that many variables affect the provision of counterparts during the project cycle and that adhering strictly to the above guidelines is sometimes difficult. If however, long-term institutional development and project sustainability is a desired objective, then this aspect of TA must be given a high priority by all parties concerned' (IBRD 1993: fn. 50).

62 This management goal is more pronounced regarding sub-Saharan Africa: 'Vague terms such as "will assist" and "will coordinate" are to be avoided as these cannot be accurately measured. Instead definitions such as "will produce a revised accounting system by (date)", "will produce a vehicle maintenance programme by (date)", etc. should be used. The key criteria here, must be the ability to measure technical assistance outputs as defined in the agreement. ... This is then accomplished by establishing a list of tasks, skills, and functions that the counterpart must be able to perform in order to replace the foreign expert. A schedule is then established for the counterpart to assume responsibility for each of the tasks and functions by specific dates' (IBRD 1994: 12).

63 On p. 210; see the technology references in Chapter 1.

64 'The Mexico project [$US48 million] had an interesting design for the privatization of research institutions, but left some important questions unanswered as to the implementation and utility of creating this sort of linkage. The Mexico project failed to achieve its major objectives of promoting real R&D activities in the private sector ...' (IBRD 1995: 9).

The 'final evaluation' of this study concludes: 'In practice, the expertise in this line of activity in the Bank is too narrowly held by a small number of specialists who have designed and implemented industrial technology development projects, starting with the first loan in Israel' (ibid.: 94). This repeats the conclusion that Forss *et al.* (1988) drew: the learning process in implementation remains limited to individuals who cannot transmit their insight to the agency (see pp. 117–122).

65 This is correct if we assume that the evaluation study took into account the conceptual innovations in the Handbook.

66 This might not be the case for the manifestations of an exo-social process, but we cannot verify this because no sector study for Chad has been published by the IBRD recently.

67 This latent process is incompatible with, for example, the Handbook's statements about the borrower overseeing the foreign consultants. Their accountability to the borrower is limited. This appears in the Handbook as an acknowledgement that local consultants provide superior accountability to the borrower. The latent process is also evident in the suggestion that using consulting companies which can exchange field personnel with home office personnel achieves a better fit of expertise (IBRD 1995: 94).

68 This can be done by deriving terms of reference (TOR) and taking into account, for example, Dambai's efforts to act upon the interface in Appui Technique. Given his narrow scope of action, excessive TOR impede, whereas suitable TOR enable him to use this scope. Assessing the interface of Autogeneración can provide TOR to help the Mexican experts gain consulting experience, something the Handbook calls for (IBRD 1993: 101).

7 Outlook

1 Not only because the intrinsic risks have to be decided by all societies; see Irwin (1995) or Doherty and Geus (1996).

2 This contributed to the political success of the Ministry of International Trade and Industry (MITI) in Japan.

3 Beyond democratic control as policy-makers believe in or pretend technological determinism, and legitimated by economists' lyrical efforts to declare these to be 'gales of creative destruction', for example. Throughout this text, we have only referred to the constructivist literature on technology (see note 5, p. 196).

4 'Whether hamlet or corporate board room, village faction or bureaucratic department, community organization or international organization, the rationale or logic which informally organizes the elaborate choreography of action and reaction remains undiscussed, inarticulate, and imperceptible to the outsider. It is not accessible to objective instrumentation; interviews, content analysis, symbolic analysis, and statistical refinements of these will invariably turn up the objective structure and symbolic logic which defend and display it' (Partridge 1987: 229).

5 Discourse analysis reveals the rhetorical habits here and there. This gap also feeds endless academic debate between defenders of alternative development, global social engineering, the White man's burden, lords of poverty and so on.

6 Instances where the observer translated were symbolically more important for the participants than the actual content of the translation, see section 3.2. Through translating, an anthropologist becomes an ally either of the developer or the developee. The fieldwork approach chosen already denied that we seek to control the gap. Ethnology as facilitator depends also on the skill of the writer, and I tried to describe analytically what developers and developees experience. Maybe another genre would have been more successful. Development anthropology's focus on rural contexts has contributed to the desire to control the gap.

7 It is possible to describe the Chadian artisans' theorizing of their work as indigenous knowledge.

8 The most important difference between the two documents is a shift from descriptive to prescriptive terminology. Instead of 'might' and 'can' (IBRD 1983), there are more examples of 'should' and 'must' for similar statements (IBRD 1993), and 'If ... then' directives instruct task managers.

9 This reflects a reduction of the affirmed TA theory by the IBRD.

10 The Chadian and French participants used technical knowledge to act on their cultural distance, something the Mexicans and US Americans could not do. This difference reflects wider social conditions, unavailable to the participants.

11 National execution or 'NEX' is a management reform in UNDP, announced in 1976 and increasingly used since 1992 (McMillan *et al.* 1997). 'To date, however, the vast majority of "mainstream" development anthropology still acts as if these dynamic project policies and policy shifts don't exist. This perspective fails to do justice to the incredible pressure these agencies are under to reform their administrative procedures and development "ideologies"' (ibid.: 321).

12 Participation is inherently limited (Pottier 1997).

13 See Juma and Ojwang (1992).

14 See Scott-Stevens (1987: 97) or Forss *et al.* (1988); as Porter *et al.* (1991: 197, 213) diagnosed, an acknowledged unpopular and hazardous position.

15 Pottier discovers this divergence from his insider perspective. In the IBRD evaluation (p. 125; and OED 1995a), the conclusion comparing Ghana and Uganda is misplaced and ascribed to IBRD management unrelated to the respective project implementation. A comparison of different TA experiences from an outsider perspective does not yield insights on implementation.

16 We presented particularly good evaluation studies as the corpus of the text because evaluations remain in the outsider perspective.

17 The institutional response to our results would then also be stronger. The actor-orientation of the fieldwork approach is a sufficient criterion to assure that,

whatever the use of research results appears to be, the cultural distance will be more actively produced.

18　The case studies cover a wide spectrum of TA institutions and contexts.

19　Writing on information management in agriculture projects, Mosse also stresses the potential of iterative and learning approaches to move beyond current planning in development institutions (Farrington *et al.* 1998: 6). We have concentrated on participant observation and neglected the conditions inside institutions because of the agency of project participants in industrial contexts, who determine the anchorage of project events in the wider social context. Latour recently proposed eleven distinct layers of sociotechnical relations (Latour 1999: 213). Each layer contains a type of interdependence between social reality and technology, and from one layer to the other the scale and entanglement changes. TA participants' agency is perhaps distinct in each layer, e.g. in agriculture social processes reflect an 'internalized ecology' (sixth layer), whereas in industry (eighth layer) social processes relevant for project implementation are more separated from technical knowledge. This would support use of the process metaphor by Farrington *et al.* for project events as well as for social processes in the context of a project in agriculture.

20　In Chad we found examples of 'reverse engineering', in which imported machines are used to develop technical knowledge. At a later stage, local educational institutions take over this function. This creates different dynamics for the latent content process and the exchange process between local and foreign participants.

21　The parties to the United Nations Framework Convention for Climate Change meet yearly; they produced the 'Kyoto Protocol' in 1997.

22　Leaping over the oil period, or, in other contexts, the charcoal or the nuclear stage.

23　Comisión Federal de Electricidad, the Mexican electricity company, and PEMEX, the oil company.

24　Sabel, C.F. (1994) 'Learning by monitoring. The institutions of economic development', in Rodwin, L. and Schön, D.A. *Rethinking the Development Experience: Essays Provoked by the Work of Albert O. Hirschman*, Washington, DC: Brookings Institution. pp. 231–274. Also Sabel, C.F. (1998) 'Panelist comments', in OED *Evaluation and Development: The Institutional Dimension*, London: Transaction Publishers, pp. 299–302.

25　Sabel refers to analytical philosophy in general and disagrees with Habermas on the applicability of the communicative/strategic acts distinction to economics (Sabel in Rodwin and Schön 1994: 270–271).

26　Bootstrapping co-operation between firms and the state in the USA often starts with professional education and research and development financing. In other countries, technology, marketing or social insurance for employees are common (Sabel 1994).

Bibliography

1 Introduction

Adler, P.S. and Cole, R.E. (1993) 'Designed for learning: a tale of two auto plants', *Sloan Management Review* Spring: 85–94.

Berger, P. (1974) *Pyramids of Sacrifice: Politics and Social Change*, New York: Basic Books.

Dennehy, R.F. and Cheng, L. (1996) 'Asian transplant management: more than Honda and Toyota', *Journal of Organizational Change Management* 9 (3): 3–5.

Freud, C. (1988) *Quelle coopération?*, Paris: Karthala.

Fry, G.W. and Thurber, C.E. (1989) *The International Education of the Development Consultant. Communicating with Peasants and Princes*, Oxford: Pergamon Press.

George, S. and Sabelli, F. (1994) *Faith and Credit*, London: Penguin.

Goulet, D. (1989) *The Uncertain Promise. Value Conflicts in Technology Transfer*, New York: New Horizons Press (first published in 1977).

Goulet, D. (1995) *Development Ethics: a Guide to Theory and Practice*, New York: Apex.

Hamnett, I. (1970) 'A social scientist among technicians', *IDS Bulletin 'Transfers, technicians and technology'* 3: 24–29.

Harris, J. (1977) 'Identification of cross-cultural talent: The empirical analysis of the Peace Corps Volunteer', in Brislin, R. (ed.) *Culture Learning: Concepts, Application and Research*, Hawaii: University of Hawaii Press.

Hawes, F. (1979) *Canadians in Development: an Empirical Study of Adaptation and Effectiveness on Overseas Assignments*, Ottawa: Canadian International Development Agency Communications Branch.

Inkster, I. (1991) *Science and Technology in History*, Basingstoke: Macmillan.

Kraar, L. (1989) 'Japan's gung-ho U.S. car plants', *Fortune* 119 (3): 78–95.

Laïdi, Z. (1989) *Enquête sur la Banque Mondiale*, Paris: Fayard.

Landes, D.S. (1998) *The Wealth and Poverty of Nations*, London: Little, Brown and Co.

Latour, B. (1987) *Science in Action, How to Follow Scientists and Engineers through Society*, Cambridge, MA: Harvard University Press.

Latour, B. (1999) *Pandora's Hope. Essays on the Reality of Science*, Cambridge, MA: Harvard University Press.

Morrill, R. (1972) 'Consultation or control? The cross-cultural advisor–advisee relationship', *Psychiatry* 35: 264–279.

Pearce, F. (1992) 'Why it's cheaper to poison the poor', *New Scientist* 133: 13.

Pinch, T., Hughes, T. and Bijker, W. (1987) *The Social Construction of Technological Systems: New Directions in the Sociology and History of Technology*, Cambridge, MA: The MIT Press.

Other publications by these authors:

Bijker, W. (1995) *Of Bicycles, Bakelites, and Bulbs: Toward a Theory of Sociotechnical Change*, Cambridge, MA: The MIT Press.

Smith, M.R. and Marx, L. (1994) *Does Technology Drive History? The Dilemma of Technological Determinism*, Cambridge, MA: The MIT Press.

Hughes, T. (1983) *Networks of Power*, Baltimore, MD: Johns Hopkins University Press.

Spitzberg, I.J. (1978) *Exchange of Expertise: the Counterpart System in the New International Order*, Boulder, CO: Westview.

Wilms, W., Hardcastle, A. and Zell, D. (1994) 'Cultural transformation at NUMMI', *Sloan Management Review* 36: 99–113.

2 Development anthropology

Achard, P. (1982) 'Sociologie du développement' ou sociologie du "développement"?', *Revue Tiers-Monde* XXIII (90): 245–256.

Achard, P. (1989) 'La passion du développement. Une analyse du discours de l'économie politique', unpublished Doctorat ès Lettres, Paris: EHESS.

Adams, A. (1986) 'An open letter to a young researcher', in Apthorpe, R. and Kráhl, A. (eds) *Development Studies: Critique and Renewal*, Leiden: E.J. Brill, pp. 218–249.

Adams, A. and Jaabe So (1996) *A Claim to Land by the River: a household in Senegal, 1720–1994*, Oxford: Oxford University Press.

Althabe, G. (1968) 'Progrès et Ostentation Economiques', *Revue Tiers-Monde* IX (33): 129–180.

Althabe, G. (1969) See the bibliography to Chapters 3 and 4.

Althabe, G. (1972) 'Les manifestations paysannes d'avril 1971', *Le Mois en Afrique – Revue française d'études politiques africaines* 78: 71–77.

Amselle, J.-L. (1990) *Logiques métisses*, Paris: Payot.

Amselle, J.-L. (1991) 'Administrateurs, développeurs et ethnologues en France: une mise en perspective historique', *Bulletin de l'APAD* 1: 17–18.

Apthorpe, R. and Gasper, D. (1996) *Arguing Development Policy: Frames and Discourses*, London: Frank Cass.

Augé, M. (1972) 'Sous-Développement et Développement: Terrain d'Etude et Objets d'Action en Afrique Francophone', *Africa* XLII (3): 205–216.

Augé, M. (1994) See the bibliography to Chapters 3 and 4.

Autumn, S. (1996) 'Anthropologists, development, and situated truth', *Human Organization* 55: 480–484.

Baré, J.-F. (1992) 'La Tunisie, la petite entreprise et la grande banque', *Cahiers des Sciences Humaines* 28: 283–304.

Baré, J.-F. (1995) *Les applications de l'anthropologie*, Paris: Karthala.

Bastide, R. (1971) *Anthropologie Appliquée*, Paris: Payot.

Bennett, J.W. and Bowen, J.R. (1988) *Production and Autonomy*, New York: University Press of America.

Berche, T. (1994) 'Un Projet de Santé en Pays Dogon', unpublished Doctorat d'Anthropologie, Paris: EHESS.

Berg, R.J. and Gordon, D.F. (1989) *Cooperation for International Development. The United States and the Third World in the 1990s*, Boulder, CO: Lynn Rienner Publishers.

Bernard, H.R. and Pelto, P.J. (1987) *Technology and Social Change*, Prospect Heights, IL: Waveland Press.

Berque, J. (1965) *De l'impérialisme à la décolonisation*, Paris: Edit. Minuit.

Boiral, P., Lanteri, J.-F. and Olivier de Sardan, J.-P. (1985) *Paysans, experts et chercheurs en Afrique noire: sciences sociales et développement rurale*, Paris: Karthala.

Bosse, H. (1979) *Diebe, Lügner, Faulenzer. Zur Ethno-Hermeneutik von Abhängigkeit und Verweigerung in der Dritten Welt*, Frankfurt: Syndikat.

Bossuyt, J., Laporte, G. and van Hoek, F. (1995) See the bibliography to Chapters 6 and 7.

Brohman, J. (1996) *Popular Development: Rethinking the Theory and Practice of Development*, Oxford: Blackwell.

Brokensha, D. and Little, P. (1988) *Anthropology of Development and Change in East Africa*, Boulder, CO: Westview.

Cernea, M. (1985) *Putting People First*, Oxford: Oxford University Press.

Cernea, M. (1995) 'Malinowski Award Lecture', *Human Organization* 54: 340–352.

Chaiken, M.S. and Fleuret, A.K. (1990) *Social Change and Applied Anthropology*, Boulder, CO: Westview.

Chambers, R. (1997) *Whose Reality Counts: Putting the First Last*, London: Intermediate Technology.

Cochrane, G. (1971) *Development Anthropology*, New York: Oxford University Press.

Cooper, F. and Packard, R. (1997) *International Development and the Social Sciences: Essays on the History and Politics of Knowledge*, Berkeley, CA: University of California Press.

Crush, J. (1995) *Power of Development*, London: Routledge.

Curry, J.C. (1996) 'Gender and lifestock in African production systems', *Human Ecology* 24 (2): 149–160.

Eddy, E. and Partridge, W.L. (1987) *Applied Anthropology in America*, New York: Columbia University Press.

Escobar, A. (1984) 'Discourse and power in development: Michel Foucault and the relevance of his work to the Third World', *Alternatives* 10: 377–400.

Escobar, A. (1991) 'Anthropology and the development encounter: the making and marketing of development anthropology', *American Ethnologist* 18: 658–682.

Escobar, A. (1995) *Encountering Development: The Making and Unmaking of the Third World*, Princeton, NJ: Princeton University Press.

Escobar, A. (1996) 'Constructing nature', in Peet, R. and Watts M. (eds) *Liberation Ecologies*, London: Routledge.

Fabian, J. (1983) *Time and the Other: How Anthropology Makes its Objects*, New York: Columbia University Press.

Fabian, J. (1989) *Power and Performance: Ethnographic Explorations through Proverbial Wisdom and Theater in Shaba, Zaïre*, Madison, WI: University of Wisconsin Press.

Fabian, J. (1991) *Time and the Work of Anthropology. Critical Essays 1971–1991*, Chur: Harwood.

Ferguson, J. (1990) *The Anti-Politics Machine: 'Development', Depolitization, and Bureaucratic Power in Lesotho*, Cambridge: Cambridge University Press (re-edited 1994).

Ferguson, J. (1997) 'Anthropology and its evil twin: "development" in the constitution of a discipline', in Cooper, F. and Packard, R. (eds) *International Development and the Social Sciences: Essays on the History and Politics of Knowledge*, Berkeley, CA: University of California Press, pp. 150–175.

Gardner, K. and Lewis, D. (1996) *Anthropology, Development and the Post-modern Challenge*, London: Pluto Press.

Gleizes, M. (1985) *Un regard sur l'ORSTOM 1943–1983: temoignage*, Paris: ORSTOM.

Gow, D. (1991) 'Collaboration in development consulting: stooges, hired guns, or musketeers?', *Human Organization* 50: 1–15.

Gow, D. (1993) 'Doubly damned: dealing with power and praxis in development anthropology', *Human Organization* 52: 380–397.

Gow, D. (1996) 'The anthropology of development: discourse, agency, and culture', *Anthropology Quarterly* 69 (3): 165–173.

Gow, D. (1997) 'Can the subaltern plan? Ethnicity and development in Cauca, Colombia', *Urban Anthropology* 26 (3–4): 243–292.

Green, E.C. (1986) *Practicing Development Anthropology*, Boulder, CO: Westview.

Grillo, R.D. and Stirrat, R.L. (1997) *Discourses of Development. Anthropological Perspectives*, Oxford: Berg.

Guichaoua, A. and Goussault, Y. (1993) *Sciences sociales et développement*, Paris: Armand Collin.

Gupta, A. and Ferguson, J. (1997) *Culture, Power, Place: Explorations in Critical Anthropology*, Durham: Duke University Press.

Guth, S. (1982) 'Exil sous contrat. Les communautés de coopérants en Afrique francophone', unpublished Doctorat ès Lettres, Paris: Université René Descartes – Paris V.

De Haan, H. and Long, N. (1997) *Images and Realities of Rural Life: Wageningen Perspectives on Rural Transformation*, Assen: Van Gorcum.

Hess, D.W. (1991) 'The development fund for Africa: new opportunities for anthropology in A.I.D.', *Studies in Third World Societies 'New directions in US foreign assistance and new roles for anthropologists'* 44: 135–146.

Hobart, M. (1993) *An Anthropological Critique of Development*, London: Routledge.

Hoben, A. (1982) 'Anthropologists and development', *Annual Review of Anthropology*, Palo Alto: Annual Reviews Inc., pp. 349–375.

Hoben, A. (1995) 'Paradigms and politics: the cultural construction of environmental policy in Ethiopia', working paper in African studies, no. 193, Boston University.

Horowitz, M. (1994) 'Development anthropology in the mid-1990s', *Development Anthropology Network* 12 (1 and 2): 1–14.

Hymes, D. (1974) *Reinventing Anthropology*, New York: Random House.

Keisman, J. (1997) 'Anthropologists are needed', WAPA newsletter, 20/4, Washington Association of Professional Anthropologists, Washington DC.

Klitgaard, R. (1990) *Tropical Gangsters*, New York: Basic Books.

Klitgaard, R. (1991) 'In search of culture: a progress report on culture and development', unpublished paper, available from the Office of the Vice President, Environmentally Sustainable Development, World Bank, Washington, DC, USA (a shortened version was published in Serageldin, I. and Taboroff, J. (1994) *Culture and Development in Africa*, Environmentally Sustainable Development Proceedings Series, no. 1, Washington, DC: The World Bank, pp. 75–120).

Klitgaard, R. (1995) 'Institutional adjustment and adjusting to institutions', World Bank discussion paper no. 303.

Koenig, D. (1997) 'Competition among Malian elites in the Manantali resettlement project: the impacts on local development', *Urban Anthropology* 26 (3–4): 369–411.

LeNaëlou, A. (1991) 'Les effets des politiques de développement à l'égard du Tiers Monde sur la construction d'une identité de la Communauté Economique Européenne (1960–1990)', unpublished Doctorat de Sociologie, Paris: EHESS.

Little, P. (1992) *The Elusive Granary: Herder, Farmer, and State in Northern Kenya*, Cambridge: Cambridge University Press.

Little, P. and Horowitz, M. (1987) *Lands at Risk in the Third World*, Boulder, CO: Westview.

Long, N. (1989) *Encounters at the Interface: A Perspective on Social Discontinuities in Rural Development*, Wageningen: The Agricultural University.

Long, N. (1994) 'The interweaving of knowledge and power in development interfaces', in Scoones, I., Thompson, J. and Chambers, R. (eds) *Beyond Farmers First*, London: Intermediate Technology Publications.

Long, N. (1996) 'Globalization and localization: new challenges to rural research', in Moore, H.L. (ed.) *The Future of Anthropological Knowledge*, London: Routledge, pp. 37–59.

Long, N. and Long, A. (1992) *Battlefields of Knowledge. The Interlocking of Theory and Practice in Social Research and Development*, London: Routledge.

Long, N. and Villarreal, M. (1998) 'Small products, big issues: value contestations and cultural identities in cross-border commodity networks', *Development and Change* 29: 725–750.

Mair, L. (1984) *Anthropology and Development*, London: Macmillan Press.

Marchand, M.H. and Parpart J.L. (1995) *Feminism/Postmodernism/Development*, London: Routledge.

Martin, M. (1998) 'The deconstruction of development: a critical overview', *Entwicklungs-ethnologie* 7: 40–59.

Mongbo, R. (1999) *The Appropriation and Dismembering of Development Interventions: Policy, Discourse, and Practice in the Field of Rural Development in Benin*, Hamburg: LIT.

Olivier de Sardan, J.-P. (1990) 'Populisme développementiste et populisme en sciences sociales: idéologie, action, connaissance', *Cahiers d'Etudes Africaines*, XXX (4/120): 475–492.

Olivier de Sardan, J.-P. (1995) *Anthropologie et Développement. Essai en socio-anthropologie du changement social*, Paris: Karthala.

ORSTOM (1996) *Sciences et Développement*, Les sciences hors d'Occident au XXe siècle series, vol. 5, Paris: ORSTOM éditions.

Partridge, W.L. (1984) *Training Manual in Development Anthropology*, Washington, DC: American Association for Anthropology and Society for Applied Anthropology.

Partridge, W.L. (1994) 'People's participation in environmental assessment in Latin America', LATEN dissemination note, no. 11, World Bank Latin America Technical Department.

Pitt, D. (1976) *Development from Below*, The Hague: Mouton.

Poncelet, M. (1994) *Une Utopie Post-Tiersmondiste*, Paris: L'Harmattan.

Pottier, J. (1993) *Practising Development: Social Science Perspectives*, London: Routledge.

Price, D. (1989) *Before the Bulldozer: The Nambiquara Indians and the World Bank*, New York: Seven Locks Press.

Rist, G. (1994) *La culture otage du développement?*, Paris: L'Harmattan.

Sabelli, F. (1993) *Recherche anthropologique et développement: éléments pour une méthode*, Paris: MSH.

Salem-Murdock, M. and Horowitz, M.M. (1990) *Anthropology and Development in North Africa and the Middle East*, Boulder, CO: Westview.

Schönhuth, M. (1991) 'The contribution of sociologists and social anthropologists to the work of development agencies', *Sonderpublikation der GTZ* no. 249.

Serageldin, I. and Taboroff, J. (1994) *Culture and Development in Africa*, Environmentally Sustainable Development Proceedings Series, no. 1, Washington, DC: The World Bank.

Sillitoe, P. (1998) 'The development of indigenous knowledge', *Current Anthropology* 39: 223–252.

Sperber, D. (1982) *Le savoir des anthropologues: trois essais*, Paris: Hermann.

Taylor, D.R.F. and Mackenzie, F. (1992) *Development From Within: Survival in Rural Africa*, London: Routledge.

van Willigen, J. (1989) *Making Our Research Useful: Case Studies in the Utilization of Anthropological Knowledge*, Boulder, CO: Westview.

van Willigen, J. (1991) *Anthropology in Use*, Boulder, CO: Westview.

Wulff, R.M. (1987) *Anthropological Praxis*, Boulder, CO: Westview.

3 and 4 Participant observation and case studies

Forss, K., Carlssen, J., Sroyland, E., Sitari, T. and Vilby, K. (1988) See the bibliography to Chapter 5.

Guth, S. (1982) See the bibliography to Chapter 2.

Hancock, G. (1989) *Lords of Poverty: the Free-wheeling Lifestyles, Power, Prestige and Corruption of the Multi-billion Dollar Aid Business*, London: Macmillan.

Kaufmann, G. (1997) 'Watching the developers', in Grillo, R.D. and Stirrat, R.L. (eds) *Discourses of Development. Anthropological Perspectives*, Oxford: Berg, pp. 107–132.

Latour, B. (1987) See the bibliography to Chapter 1.

Latour, B. (1999) See the bibliography to Chapter 1.

Pinch, T., Hughes, T. and Bijker, W. (1987) See the bibliography to Chapter 1.

Scott-Stevens, S. (1987) See the bibliography to Chapters 6 and 7.

Talalay, M., Farrands, C. and Tooze, R. (1997) *Technology, Culture and Competitiveness*, London: Routledge.

Todorov, T. (1984) See the bibliography to Chapters 6 and 7.

French contemporary anthropology

Abélès, M. (1991) *Quiet Days in Burgundy: a Study of Local Politics*, Cambridge: Cambridge University Press.

Abélès, M. (1992) *La Vie quotidienne au Parlament Européen*, Paris: Hachette.

Abélès, M. and Jeudy, H.-P. (1997) *Anthropologie du politique*, Paris: A. Colin.

Althabe, G. (1968) See the bibliography to Chapter 2.

Althabe, G. (1969) *Oppression et libération dans l'imaginaire*, Paris: Maspero.

Althabe, G. (1972) See the bibliography to Chapter 2.

Althabe, G. (1990a) 'Ethnologie du contemporain et enquête de terrain', *Terrain* 'Ethnologie Urbaine', 14: 126–131.

Althabe, G. (1990b) 'L'ethnologue et sa discipline', *L'Homme et la Société*, XXIV (95–96): 25–41.

Althabe, G. and Sélim, M. (1998) *Démarches ethnologiques au présent*, Paris: L'Harmattan.

Althabe, G., Marcadet, C., de la Pradelle, M. and Sélim, M. (1984) *urbanisme et enjeux quotidiens*, Paris: éditions anthropos.

Althabe, G., Fabre, M. and Lenclud, G. (1992) *Vers une ethnologie du présent*, Paris: MSH.

Augé, M. (1994) *Pour une anthropologie des mondes contemporains*, Paris: Aubier.

Augé, M. (1995) *Non-places: Introduction to an Anthropology of Supermodernity*, London: Verso.

Augé, M. (1998) *A Sense for the Other: the Timeliness and Relevance of Anthropology*, Stanford, CA: Stanford University Press.

Augé, M. (1999) *An Anthropology for Contemporaneous Worlds*, Stanford, CA: Stanford University Press.

Friedman, J. (all dates) See the biblography to Chapter 5.

Terray, E. (1972) *Marxism and 'Primitive' Societies: Two Studies*, New York: Monthly Review Press.

Terray, E. (1990) *La politique dans la caverne*, Paris: Seuil.

Terray, E. (1994) *Une passion allemande: Luther, Kant, Schiller, Hölderlin, Kleist*, Paris: Seuil.

German development ethnology

Antweiler, C. (1987) *Ethnologische Beiträge zur Entwicklungspolitik 1*, Bonn: Politischer Arbeitskreis Schulen.

Bierschenk, T. and Elwert, G. (1993) *Entwicklungshilfe und ihre Folgen: Ergebnisse empirischer Untersuchungen*, Frankfurt: Campus.

Bliss, F. and Neumann, S. (1996) *Ethnologische Beiträge zur Entwicklungspolitik 3*, Bonn: Politischer Arbeitskreis Schulen.

Bliss, F. and Schönhuth, M. (1990) *Ethnologische Beiträge zur Entwicklungspolitik 2*, Bonn: Politischer Arbeitskreis Schulen.

Grammig, T. (1992) 'Les formateurs à l'école de l'apprentissage', *Enseignement Technique et Développement Industriel* 4: 24–26 (GRET, Paris; see Appendix 1).

Grammig, T. (1993) 'Unterstützende Empirie in der Projektimplementierung', *Entwicklungsethnologie* (1/2): 21–37.

Grammig, T. (1996) 'Ethnologie und Entwicklung', *Entwicklungsethnologie* 5 (2): 96–106.

Kievelitz, U. (1988) *Kultur, Entwicklung und die Rolle der Ethnologie*, Bonn: Politischer Arbeitskreis Schulen.

Moser-Schmitt, E. (1984) 'Ein ethnologischer Beitrag zum Bereich Communication and Community Developement in einem nepalesischen Stadtentwicklungsprojekt', *Zeitschrift für Ethnologie* 109: 125–141.

Schönhuth, M. (1998) 'Entwicklungsethnologie in Deutschland. Eine Bestandsaufnahme aus Sicht der Arbeitsgemeinschaft Entwicklungsethnologie und ein Vergleich mit internationalen Entwicklungen', *Entwicklungsethnologie* 7: 11–39.

Schönhuth, M. and Kievelitz, U. (1991) See the bibliography to Chapter 2.

Schönhuth, M. and Kievelitz, U. (1994) *Participatory Learning Approaches, Rapid Rural Appraisal, Participatory Appraisal, an Introductory Guide*, Rossdorf: TZ-Verlaggesellschaft.

Local environment

Mexico

Cypess, S.M. (1991) *La Malinche in Mexican Literature: from History to Myth*, Austin, TX: University of Texas Press.

Hagnes, K. (1991) 'Dependency, postimperialism, and the Mexican revolution: an historiographic review', *Mexican Studies/Estudios Mexicanos* 7 (2): 225–251.

Monteforte, R. (1992) 'A review of power sector economics in Mexico', *The Journal of Energy and Development* 16: 15–36.

Nunez Becerra, F. (1996) *La Malinche: de la Historia al Mito*, Mexico DF: Instituto Nacional de Antropologia e Historia.

Paz, O. (1961) *The Labyrinth of Solitude: Life and Thought in Mexico*, New York: Grove Press.

Paz, O. (1972) 'Eroticism and gastrosophy', *Daedalus 'How Others See the United States'* 101 (4): 67–85.

Chad

Abdelsalam, C. (1991) 'Les Strategies des ONG au Tchad: de l'urgence au développement', unpublished Diplom d'Etude Approfondie, Paris: INALCO.

Azevedo, M.J. and Nnadozie, E.U. (1998) *Chad: A Nation in Search of Its Future*, Boulder, CO: Westview.

Bouquet, C. (1982) *Tchad, genèse d'un conflit*, Paris: L'Harmattan.

Buijtenhuijs, R. (1985) 'La mort du commandant Galopin, une mise au point', *Politique Africaine* December: 91–95.

Buijtenhuijs, R. (1993) *La conférence nationale souveraine du Tchad*, Paris: Karthala.

Foures, A. (1986) *Au-dela du sanctuaire*, Paris: Economica.

Général Ingold (1945) *L'Epopée Leclerc au Sahara*, Paris: Berger-Levrault.

Harre Igué and Arditi, C. (1990) *Les échanges marchands entre le Tchad, le Nord Nigeria et le Nord Cameroun*, Paris: L'Harmattan.

Labazee, P. (1988) *Entreprises et entrepreneurs du Burkina Faso*, Paris: Karthala.

Rabier, Y. (1993) 'Politique Internationale du Conflit Tchadien (1960–1990) guerre civile et système mondial', unpublished Doctorat d'Etat en Science Politique à l'Institut d'Etudes Politiques de Paris.

de Raulin, A. (1990) 'Les ONG au Tchad', *Revue juridique et politique* 44 (3): 440–459.

Tubiana, J. (1994) *L'identité Tchadienne*, Paris: L'Harmattan.

Warnier, J.-P. (1993) *Les Entreprises au Cameroun*, Paris: Karthala.

5 Latent processes

Alavi, H. (1973) 'Peasant classes and primordial loyalties', *Journal of Peasant Studies* 1: 22–62.

Atmanspacher, H. and Dalenoort, G.J. (1994) *Inside Versus Outside: Endo- and Exo-Concepts of Observation and Knowledge in Physics, Philosophy and Cognitive Science*, Berlin: Springer.

Baré, J.-F. (1998) 'Of loans and results: elements for a chronicle of evaluation at the World Bank', *Human Organization* 57 (3): 319–325.

Bierschenk, T. (1990) 'Planspiel zur entwicklungspolitischen Projektpraxis. Ein Beitrag zur Entdämonisierung der Entwicklungshilfe', *Sozialanthropologische Arbeitspapiere*, no. 23, Berlin: Das arabische Buch.

Bohman, J.F. (1986) 'Formal pragmatics and social criticism: the philosophy of language and the critique of ideology in Habermas' Theory of Communicative Action', *Philosophy and social critique* 11 (4): 331–353.

Breckenridge, C.A. (1993) See the bibliography to Chapters 6 and 7.

Carlsson, J., Köhlin, G. and Ekbom, A. (1994) *The Political Economy of Evaluation. International Aid Agencies and the Effectiveness of Aid*, New York: St. Martin's Press.

Carlsson, J., Forss, K., Metell, K., Segestam, L. and Strömberg, T. (1997) *Using the Evaluation Tool. A Survey of Conventional Wisdom and Common Practice at SIDA*, Stockholm: SIDA Department for Evaluation and Internal Audit.

Ekholm, K. and Friedman, J. (1995) 'Global complexity and the simplicity of everyday life', in Miller, D. (ed.) *Worlds Apart. Modernity Through the Prism of the Local*, London: Routledge, pp. 134–168.

Ellul, J. (1990) *The Technological Bluff*, Grand Rapids, MI: Eerdmans.

Feenberg, A. (1991) *Critical Theory and Technology*, Oxford: Oxford University Press.

Feenberg, A. (1992) 'Subversive rationalization: technology, power and democracy', *Inquiry* 35 (3/4): 301–322.

Feenberg, A. (1995) *Alternative Modernity: the Technical Turn in Philosophy and Social Theory*, Berkeley, CA: University of California Press.

Forss, K. (1985) *Planning and Evaluation in Aid Organizations*, Stockholm: The Economic Research Institute Stockholm School of Economics.

Forss, K. and Carlsson J. (1997) 'The quest for quality – or can evaluation findings be trusted', *Evaluation* 3/4: 481–501.

Forss, K., Carlssen, J., Sroyland, E., Sitari, T. and Vilby, K. (1988) *The Effectiveness of Technical Assistance Personnel*, Copenhagen: DANIDA.

Friedman, J. (1990) 'Being in the world: localization and globalization', in Featherstone, M. (ed.) *Global Cultures*, London: Sage.

Friedman, J. (1992a) 'Narcissism, roots and postmodernity: the construction of selfhood in the global crisis', in Friedman, J. and Scott, L. (eds) *Modernity and Identity*, Oxford: Blackwell, pp. 331–366.

Friedman, J. (1992b) 'The past in the future', *American Anthropologist* 94 (4): 837–859.

Friedman, J. (1992c) 'Myth, history and political identity', *Cultural Anthropology* 7: 194–210.

Friedman, J. (1994) *Cultural Identity and Global Process*, London: Sage.

Friedman, J. (forthcoming) *Open Systems and Closed Minds*.

Galtung, J. (1978) *Whither Technical Assistance? On the Future of International Development Co-operation*, Geneva: IUED.

Galtung, J. (1979) *Development, Environment and Technology. Towards a Technology for Self-reliance*, Geneva: UN Conference on Trade and Development.

Giddens, A. (1979) *Central Problems in Social Theory: Action, Structure and Contradiction in Social Analysis*, London: Macmillan.

Giddens, A. (1984) *The Constitution of Society: Outline of the Theory of Structuration*, Cambridge: Polity Press.

Giddens, A. (1987) *Social Theory and Modern Sociology*, Oxford: Polity Press.

Gjerding, A.N. (1998) 'Technical collaboration as social construction', *International Business Economics Working Paper Series*, no. 29, Aalborg: University of Aalborg.

Guth, S. (1982) See the bibliography to Chapter 2.

Habermas, J. (1968) *Antworten auf Herbert Marcuse*, Frankfurt: Suhrkamp.

Habermas, J. (1970) *Toward a Rational Society*, Boston: Beacon Press (includes the translation of *Technik und Wissenschaft als Ideologie*, Frankfurt: Suhrkamp).

Habermas, J. (1984) *The Theory of Communicative Action*, Boston: Beacon Press.

Habermas, J. (1988) *On the Logic of the Social Sciences*, Cambridge: Polity Press.

Habermas, J. (1992) *Postmetaphysical Thinking: Philosophical Essays*, Cambridge, MA: The MIT Press.

Habermas, J. (1998) *On the Pragmatics of Communication*, Cambridge, MA: The MIT Press.

Hancock, G. (1991) *Lords of Poverty: the Power, Prestige and Corruption of the International Aid Business*, New York: Atlantic Monthly Press.

Hawes, F. (1979) See the bibliography to Chapter 1.

Heller, P.B. (1985) *Technology Transfer and Human Values. Concepts, Applications, Cases*, Lanham: University Press of America.

Hindness, B. (1986) 'Actors and social relations', in Wardell, M.L. and Turner, S.P. (eds.) *Sociological theory in transition*, Boston, MA: Allen and Unwin.

Hirschman, A.O. (1965) 'Obstacles to development: a classification and a quasi-vanishing act', *Economic Development and Cultural Change* XIII (4): 385–393.

Hirschman, A.O. (1967) *Development Projects Observed*, Washington, DC: Brookings Institutions.

Hirschman, A.O. (1976) 'On Hegel, imperialism, and structural stagnation', *Journal of Development Economics* 3: 1–8.

Hirschman, A.O. (1977) 'Foreign aid: a critique', in Stout, R. (ed.) *Readings in Project Design*, Bloomington, IN: USAID and Midwest Universities Consortium for International Activities.

Hirschman, A.O. (1987) 'The search for paradigms as a hindrance to understanding', in Rabinow, P. and Sullivan, W.M. (eds) *Interpretative Social Science: A Second Look*, Berkeley, CA: University of California Press.

Hirschman, A.O. (1995) *A Propensity for Self-Subversion*, Cambridge, MA: Harvard University Press.

Hoebink, P. (1995) *The Comparative Effectiveness and Evaluation Efforts of EU Donors*, The Hague: National Advisory Council for Development Cooperation.

IBRD (1993) See the bibliography to Chapters 6 and 7.

Kapferer, B. (1976) *Transaction and Meaning: Directions in the Anthropology of Exchange Symbolic Behaviour*, Philadelphia, PA: Institute for the Study of Human Issues.

Kaponen, J. and Mattila-Wiro, P. (1996) *Effects or Impacts? Synthesis Study on Evaluations and Reviews Commissioned by Finnida 1988 to mid-1995*, Helsinki: Department for International Development Cooperation.

Knorr Cetina, K. (1997) 'Sociality with objects: social relations in postsocial knowledge societies', *Theory, Culture and Society* 14 (4): 1–30.

Long, N. (1989) See the bibliography to Chapter 2.

Marcuse, H. (1964) *One-dimensional Man: Studies in the Ideology of Advanced Industrial Society*, London: Routledge.

Mbembe, A. (forthcoming) *Post-colonial Relations*.

McNaughton, P.R. (1988) *The Mande Blacksmiths: Knowledge, Power and Art in West Africa*, Bloomington, IN: Indiana University Press.

OECD/DAC (1984) *Report of the Expert Group on Aid Evaluation on Lessons of Experience Emerging from AID evaluations*, Paris: OECD.

OED (1994) 'World Bank relations with Mexico', *OED Précis*, no.71, Washington, DC: IBRD Operations Evaluation Department.

OED (1995a) *Evaluation and Development. Proceedings of the 1994 World Bank Conference*, Washington, DC: IBRD Operations Evaluation Department.

OED (1995b) 'Designing technical assistance projects: lessons from Ghana and Uganda', *OED Précis*, no. 95, Washington, DC: IBRD Operations Evaluation Department.

OED (1996) 'Technical assistance', *Lessons and Practices*, no.7, Washington, DC: IBRD Operations Evaluation Department.

OED (1998) *Evaluation and Development: The Institutional Dimension*, London: Transaction Publishers.

ORSTOM (1986) *L'Exercice du Développement*, Paris: Editions de l'ORSTOM.

Podolny, J.M. and Stuart T.E. (1995) 'A role-based ecology of technological change', *American Journal of Sociology* 100 (5): 1224–1260.

Poncelet, M. (1994) See the bibliography to Chapter 2.

Rebien, C.C. (1996) *Evaluating Development Assistance in Theory and in Practice*, Aldershot: Avebury.

Rodwin, L. and Schön, D.A. (1994) *Rethinking the Development Experience: Essays Provoked by the Work of Albert O. Hirschman*, Washington, DC: Brookings Institution.

Simondon, I.J. (1989) *Le mode d'existence des objets techniques*, Paris: Aubier.

Stocking, G.W. (1991) *Colonial Situations. Essays on the Contextualization of Ethnographic Knowledge*, Madison, WI: University of Wisconsin Press.

Thomas, R.J. (1994) *What Machines Can't Do: Politics and Technology in the Industrial Enterprise*, Berkeley, CA: University of California Press.

Van Velzen, H.U.E. (1973) 'Robinson Crusoe and Friday: strength and weakness of the big man paradigm', *MAN* 8: 592–612.

Wood, D.H., Wood, J.M., Turner, A. and Kean, J. (1988) 'Synthesis of A.I.D. Evaluation Reports: FY 1985 and FY 1986', *A.I.D. Evaluation Occasional Paper*, no.16.

Zimmerman, M.E. (1979) 'Heidegger and Marcuse: technology as ideology', *Research in Philosophy and Technology* 2: 245–261.

Zimmerman, M.E. (1990) *Heidegger's Confrontation with Modernity. Technology, Politics and Art*, Bloomington, IN: Indiana University Press.

6 and 7 Event management and outlook

Argyris, C. (1978) *The Learning Organization*, Reading, MA: Addison-Wesley.

Berg, E.J. (1993) *Rethinking Technical Cooperation. Reforms for Capacity Building in Africa*, New York: UNDP Regional Bureau for Africa and Development Alternatives Inc.

Black, J.K. (1991) *Development in Theory and Practice. Bridging the gap*, Boulder, CO: Westview.

Bossuyt, J., Laporte G. and van Hoek F. (1995) 'New avenues for TC in Africa. Improving the record in terms of capacity building', *Policy Management Report*, no. 2, Maastricht: European Centre for Development Policy Management.

Breckenridge, C.A. and van der Veer, P. (1993) *Orientalism and the Postcolonial Predicament*, Philadelphia, PA: University of Pennsylvania Press.

Castoriadis, C. (1987) *The Imaginary Institution of Society*, Cambridge: Polity Press.

Chauveau, J.-P. (1992) 'Du populisme bureaucratique dans l'histoire institutionnelle du développement rural en Afrique de l'Ouest', *Bulletin de l'APAD* 4: 23–32.

Coombs, R., Richards, A., Saviotti, P. and Walsh, V. (1996) *Technological Collaboration*, Cheltenham: Edward Elgar Publishing.

Cusworth, J.W. and Franks, T.R. (1993) *Managing Projects in Developing Countries*, Harlow: Addison Wesley Longman.

Doherty, B. and de Geus, M. (1996) *Democracy and Green Political Thought: Sustainability, Rights and Citizenship*, London: Routledge.

Farrington, J., Mosse, D. and Rew, A. (1998) *Development as Process: Concepts and Methods for Working with Complexity*, London: Routledge.

Ferguson, J. (1990) See the bibliography to Chapter 2.

Forss, K., Carlssen, J., Sroyland, E., Sitari, T. and Vilby, K. (1988) See the bibliography to Chapter 5.

Friedman, J. (1992a) See the bibliography to Chapter 5.

Fry, G.W. and Thurber, C.E. (1989) See the bibliography to Chapter 1.

Général Ingold (1945) See the bibliography to Chapters 3 and 4.

Glaser, W.A. (1975) 'Making better use of technical assistance experts', *Focus: Technical Cooperation*, quarterly supplement to the *International Development Review* 4 and 2: 21–25.

Gow, D. and Morss, E.R. (1985) *Implementing Rural Development Projects. Lessons from AID and World Bank Experiences*, Boulder, CO: Westview.

Gow, D. and Morss, E.R. (1988) 'The notorious nine: critical problems in project implementation', *World Development* 16: 1399–1418.

Habermas, J. (1988) See the bibliography to Chapter 5.

Hagnes, K. (1991) See the bibliography to Chapters 3 and 4.

Hirschman, A.O. (1967) See the bibliography to Chapter 5.

Hirschman, A.O. (1977) See the bibliography to Chapter 5.

Hoben, A. (1989) 'USAID: organizational and institutional issues and effectiveness', in Berg, R.J. and Gordon, D.F. (eds) *Cooperation for International Development. The United States and the Third World in the 1990s*, Boulder, CO: Lynn Rienner Publishers, pp. 253–278.

Hoben, A. (1991) 'The integrative revolution revisited', *World Development* 19: 17–30.

Honadle, G. and Cooper, L. (1990) 'Closing the loops: workshop approaches to evaluating development projects', in Finsterbush, K., Ingersoll, J. and Llewellyn, L. (eds) *Methods for Social Analysis in Developing Countries*, Boulder, CO: Westview.

IADB (1988) *Technical Cooperation for Institutional Strengthening. An Evaluation of the Bank's Experience*, Washington, DC: IADB Operations Evaluation Office.

IBRD (1983) 'Managing project-related technical assistance. The lessons of success', *World Bank Staff Working Paper*, no. 586, Washington, DC: IBRD.

IBRD (1992) *Effective Implementation: Key to Development Impact. Portfolio Management Task Force Report*, Washington, DC: Internal World Bank Document (Wapenhans Report).

IBRD (1993) *Handbook on Technical Assistance*, Washington DC: IBRD Operations Policy Department.

IBRD (1994) 'Technical assistance in Africa: how it works, and doesn't work', *AFTHR Technical Note*, no. 16, Washington, DC: IBRD Human Resources and Poverty Division.

IBRD (1995) *Developing Industrial Technology. Lessons for Policy and Practice*, Washington, DC: IBRD – A World Bank Operations Evaluation Study.

IRAM (1998) *Regards Du Sud. Des sociétés qui bougent, une coopération à refonder*, Paris: L'Harmattan.

Irwin, A. (1995) *Citizen Science: a Study of People, Expertise and Sustainable Development*, London: Routledge.

Juma, C. and Ojwang, J.B. (1992) *Technology Transfer and Sustainable Development: International Policy Issues*, Nairobi: ACTS Press.

Kaplan, A. (1996) *Development Practitioners' Handbook*, London: Pluto.

Kaplinsky, R. (1994) *Easternisation: The Spread of Japanese Management Techniques to Developing Countries*, London: Frank Cass.

Keynes, J.M. (1936) *The General Theory of Employment, Interest and Money, The Collected Writings of John Maynard Keynes Vol. VII*, London: Macmillan.

Latour, B. (1999) See the bibliography to Chapter 1.

McMillan, D.E., Jallow, O. and Mulder, H. (1997) 'Renegotiating development partnerships: a case study of national execution of a UNDP program in The Gambia', *Urban Anthropology* 26 (3–4): 293–329.

OED (1995a) See the bibliography to Chapter 5.

OED (1998) See the bibliography to Chapter 5.

Partridge, W.L. (1987) 'Towards a theory of practice', in Eddy, E. and Partridge, W.L. (eds) *Applied Anthropology in America*, New York: Columbia University Press, pp. 211–233.

Paul, S. (1983) *Managing Development Programmes: the Lessons of Success*, Boulder, CO: Westview.

Paulais, T. (1995) *Le développement urbain en Côte d'Ivoire*, Paris: Karthala.

Paz, O. (1961) See the bibliography to Chapters 3 and 4.

Paz, O. (1972) See the bibliography to Chapters 3 and 4.

Piore, M. and Sabel, C.F. (1984) *The Second Industrial Divide: Possibilities for Prosperity*, New York: Basic Books.

Poncelet, M. (1994) See the bibliography to Chapter 2.

Porter, D., Allen, B. and Thompson, G. (1991) *Development in Practice. Paved with Good Intentions*, London: Routledge.

Pottier, J. (1993) See the bibliography to Chapter 2.

Pottier, J. (1997) 'Towards an ethnography of participatory appraisal and research', in Grillo, R.D. and Stirrat, R.L. (eds) *Discourses of Development. Anthropological Perspectives*, Oxford: Berg, pp. 203–228.

Rabier, Y. (1993) See the bibliography to Chapters 3 and 4.

Reineke, R.-D. and Sülzer R. (1995) *Organisationsberatung in Entwicklungsländern. Konzepte und Fallstudien*, Wiesbaden: Gabler.

Rist, G. (1994) See the bibliography to Chapter 2.

Rodwin, L. and Schön D.A. (1994) See the bibliography to Chapter 5.

Rondinelli, D.A. (1987) *Development Administration and U.S. Foreign Aid Policy*, Boulder, CO: Lynne Rienner Publishers.

Rondinelli, D.A. (1989) 'Reforming U.S. foreign aid policy: constraints on development assistance', *The Policy Studies Journal* 18: 67–85.

Rondinelli, D.A. (1993) *Development Projects as Policy Experiments: an Adaptive Approach to Development Administration*, 2nd edn, London: Routledge.

Rosenberg, N. (1982) *Inside the Black Box: Technology and Economics*, New York: Cambridge University Press.

Sabel, C.F. (1993) 'Studied trust: building new forms of cooperation in a volatile economy', *Human Relations* 46 (9): 1133–1170.

Sabel, C.F. (1994) 'Bootstrapping reform: rebuilding firms, the welfare state and unions', *Politics and Society* 23: 5–48.

Sabel, C.F. and Zeitlin, J. (1997) *World of Possibilities: Flexibility and Mass Production in Western Industrialization*, Cambridge: Cambridge University Press.

Sahara, T. (1990) 'A comparative study of approaches to institution building by aid agencies', unpublished M.A. Thesis, University of Manchester: IDPM.

Schein, E.H. (1993) *Organizational Culture and Leadership*, San Francisco: Jossey-Bass.

Scott-Stevens, S. (1987) *Foreign Consultants and Counterparts: Problems in Technology Transfer*, Boulder, CO: Westview.

Senge, P.M. (1990) *The Fifth Discipline: The Art and Practice of the Learning Organization*, New York: Doubleday.

Spencer, D.L. (1970) *Technology Gap in Perspective: Strategy of International Technology Transfer*, New York: Spartan Books.

Stokke, O. (1996) *Foreign Aid Towards the Year 2000: Experiences and Challenges*, London: Frank Cass.

Sunshine, R.B. (1990) *Negotiating for International Development: a Practitioner's Handbook*, Dordrecht: Martinus Nijhoff Publishers.

Sunshine, R.B. (1995) *Managing Technical Assistance: a Practitioners' Handbook*, Honolulu: East–West Center.

Todorov, T. (1984) *The Conquest of America: the Question of the Other*, New York: Harper & Row.

Todorov, T. (1986) 'Le croisement des cultures', *Communications* 43: 5–24.

Todorov, T. (1988) 'Knowledge in social anthropology: distancing and universality', *Anthropology Today* 4 (2): 2–5.

Todorov, T. (1990) *Genres in Discourse*, Cambridge: Cambridge University Press.

Todorov, T. (1993) *On Human Diversity: Nationalism, Racism, Exoticism in French Thought*, Cambridge, MA: Harvard University Press.

Todorov, T. (1995a) *The Morals of History*, Minneapolis, MN: University of Minneapolis Press.

Todorov, T. (1995b) *La vie commune: essai d'anthropology generale*, Paris: Seuil.

Index